LARISA   ALTSHULER

ARE   YOU   GIFTED   IN   MATH ?

A BOOK FOR THE JOY OF CHALLENGE

"...Mathematics, like music, is worth doing for its own sake."

"Just as appreciation of music is a hallmark of the educated person, so should be an appreciation of mathematics."

Peter Hilton

Distinguished Professor of Mathematics, Emeritus

State University of New York at Binghamton

To all my students who share with me the love of Mathematics

PREFACE

The idea to write this book came to my mind during my involvement in preparing challenging website problems for high school children and undergraduate students interested in Mathematics.

Teaching mathematics is an exciting and challenging process. I changed my profession twice before I recognized that I wanted to teach mathematics. I worked as an instructor for pilots and mechanics, teaching them the constructions of airplanes and engines. I was also a research mathematician at the Institute for Low Temperature. The idea of becoming a math teacher ripened in me for a long time until finally, at the age of thirty nine, to the horror of my family, I had decided to restart my career and accepted a position of a high school math teacher. For more than three decades, I taught mathematics in high schools and universities throughout Ukraine, Lithuania, and USA. I have never regretted my decision.

My teaching credo is based on my deep belief that the vast majority of students have the ability not only to learn mathematics but also to enjoy the overall process. Among many goals of teaching math I would like to emphasize the following:

* "Systematically encourage students to think independently" (George Polya);
* Show the beauty of the subject;
* Help each student to discover the ability to be a winner;
* Help every student to get skills necessary to learn other sciences;
* Convince pupils that math is not only "food for the mind" but also as pleasurable as listening to your favorite music while still being practical in life matters;
* Discover that someone is gifted in math, develop her/his talent and have mutual joy of such discovery.

To reach these goals in each mathematical topic, a teacher has to have a collection of interesting, diverse, and challenging problems with different levels of difficulty. And here is the book which has been created for boys and girls, their parents and teachers, young and old, for those who already love Math and for those who have yet to discover it. The 250 problems include a variety of topics – Numbers and Sequences, Clock Arithmetic, Polynomials, Equations and Inequalities, Geometry, Trigonometry, Functions, and Story Problems - and many levels of difficulty, from quite basic to what might appear on a mathematical competition. The majority are unconventional according to the standards of most high school math programs. Since most problems do require a basic knowledge of high school math, pointers are included to refresh your memory about the necessary facts and formulas. Whatever more you need, you will find in our introduction to the topic. Make note of the fact that the problems are NOT arranged in the order of difficulty; you can find simpler problems in the end of the book and not in the beginning.

I would be truly happy if this book would help change someone's attitude from "I hate Math" to "I love Math" or recognize that you are gifted in Mathematics. What might be better than such a reward for a teacher?

ACKNOWLEDGMENTS

I will always remember my first teacher of mathematics. In 1945, fourty of 7-year old students including me, dressed in coats and gloves, sat in a cold unheated room in the destructed by war city of Kharkov, Ukraine. In such dreadful conditions, my teacher introduced me to the beautiful world of Mathematics. She involved me in the exciting process of thinking, and I discovered the unforgettable sensation: " I can do it! I can be a winner in the struggle with the Math problem!" Thank you Vera Matveevna Samohvalova!

At the end of 1980s, I was a high school teacher in Klaipeda, Lithuania. Under my supervision, ten gifted in Math students overcame a complete three-year course of mathematics in Corresponding Mathematics School at Leningrad State University. All my students successfully graduated from this school and were enrolled in the prestigious universities in Russia, Lithuania, and Ukraine. I always remember those years of mutual joy of teaching and learning Mathematics with sincere gratitude to all my students. I would like to thank Oksana Kruglikova, Andrew Kostenkov, Max Kursenev, Andrew Noskov, Paul Kirgner, Sergey Babitskij, Audra and Aushra Shikshniute for their enthusiasm, love of Math, and our wonderful relationships. I would like to extend my appreciation to two other talented Math students of the same school, Igor Popov and Vitaliy Smetanin, whose genuine interest in Math inspired me for many years.

I would like to thank my colleagues in Math Department of Kent State University, where I taught a variety of undergraduate courses of Math. Their heartwarming support, tolerance and friendly environment made it possible for me to write this book. The discussions with them about the problems teachers encounter in many American schools prompted me to use my own experience in creating material for high school teachers to attract their pupils to Math. I would like to give many thanks to my colleague Mary Beth Rollick whose sincere interest to my work helped me to select the intriguing problems for this book.

I would like to express my very special appreciation to professor David Presser for editing the text, making this book legible for an American audience.

Finally, to my beloved husband Vladimir and my children Victoria and Eugene: "Thank you so much! Your support made creating of this book enjoyable."

NOTATIONS

In this book we adopted non-traditional, but in our opinion more convenient notation designed to stress intuitive comprehension over formal correctness. For example, a segment or an angle and its measure are denoted identically. If we write $AB = a$, we mean that the length of the segment AB is equal to "a" units. Analogically, $\angle ABC = 30°$ means that the measure of the angle ABC is equal to $30°$. Our teaching experience convince us that such notations are completely understandable in context , and can never be a reason for misinterpretation of information about geometric figures.

CONTENTS

NUMBERS, SEQUENCES , POLYNOMIALS

Let us remind ourselves of popular formulas often used to solve various problems in different parts of Mathematics.

I. SPECIAL PRODUCT FORMULAS

1. $(a + b)(a - b) = a^2 - b^2$ (1)
2. $(a \pm b)^2 = a^2 \pm 2ab + b^2$ (2)
3. $(a \pm b)^3 = a^3 \pm 3a^2b + 3ab^2 \pm b^3$ (3)

Observing the last two formulas we come to the question : Does there exist a formula that gives the expansion of $(a + b)^n$ for any natural number n ? The answer was given by socalled Binomial Theorem proved by one of the giants of physics and mathematics, Sir Isaac Newton. But before we state his formula we should turn back to the discovery of the another great physicist and mathematician, Blaise Pascal. It is interesting that the triangle, formed by certain special numbers, was known earlier and published in a Chinese document by Chu Shi-Kie dated 1303, so Pascal rediscovered this triangle.

Pascal's Triangle has the following form:

$$
\begin{array}{ccccccccccc}
 & & & & & 1 & & & & & \\
 & & & & 1 & & 2 & & 1 & & \\
 & & & 1 & & 3 & & 3 & & 1 & \\
 & & 1 & & 4 & & 6 & & 4 & & 1 \\
 & 1 & & 5 & & 10 & & 10 & & 5 & & 1 \\
\end{array}
$$

....

Notice that in each row the first and last number is 1, and every other number is located between the numbers of the preceding row and is equal to the sum of those two numbers. The terms of each row of Pascal's Triangle are the coefficients in the expansion of $(a + b)^n$, where n is the number of the row.

For example $(a + b)^5 = a^5 + 5a^4b + 10a^3b^2 + 10a^2b^3 + 5ab^4 + b^5$.

So using Pascal's Triangle we can find the binomial expansion for any n, but this is not practical for large values of n because we always have to know the coefficients of the preceding row.
Now we will give Newton's Formula, which allows us to find the expansion of $(a + b)^n$ for any natural number n:

$$(a \pm b)^n = C_n^0 a^n \pm C_n^1 a^{n-1} b^1 + C_n^2 a^{n-2} b^2 \pm C_n^3 a^{n-3} b^3 + ... \pm C_n^n b^n \qquad (4)$$

where $C_n^k = \dfrac{n!}{k!(n - k)!}$ ($n! = 1 \cdot 2 \cdot 3 \cdot 4 \cdot ... \cdot n, \quad 0! = 1$)

Example : $(a + b)^5 = C_5^0 a^5 + C_5^1 a^4 b^1 + C_5^2 a^3 b^2 + C_5^3 a^2 b^3 + C_5^4 a^1 b^4 + C_5^5 b^5$ where

$$C_5^0 = \frac{5!}{0!(5-0)!} = \frac{1 \cdot 2 \cdot 3 \cdot 4 \cdot 5}{1 \cdot 1 \cdot 2 \cdot 3 \cdot 4 \cdot 5} = 1, \quad C_5^1 = \frac{5!}{1!(5-1)!} = 5, \quad C_5^2 = \frac{5!}{2!3!} = 10, \quad C_5^3 = \frac{5!}{3!2!} = 10,$$

$$C_5^4 = \frac{5!}{4!1!} = 5, \quad C_5^5 = \frac{5!}{5!0!} = 1.$$

II. FACTORING POLYNOMIALS

Reading formulas (1) and (2) from right to left, we have formulas for factoring a difference of squares and a perfect square of a binomial. Let us add here a popular formula for factoring the sum or difference of cubes:

$$a^3 \pm b^3 = (a \pm b)(a^2 \mp ab + b^2) \tag{5}$$

The next obvious question we can ask is : Does there exist a formula for factoring a binomial $a^n \pm b^n$ for any natural number n ?
The formula for $a^n - b^n$ can be determined by direct multiplication:

$$a^n - b^n = (a - b)(a^{n-1} + a^{n-2}b + a^{n-3}b^2 + \ldots + b^{n-1}) \tag{6}$$

Conclusion: $a^n - b^n$ is divisible by $a - b$.
If n is odd, then by substituting b for -b in the previous formula we have

$$a^n + b^n = (a + b)(a^{n-1} - a^{n-2}b + a^{n-3}b^2 - \ldots + b^{n-1}) \tag{7}$$

Conclusion: $a^n + b^n$ is divisible by $a + b$ for any odd natural number n .

III. SEQUENCES

A sequence is a set of the numbered terms $a_1, a_2, a_3, \ldots, a_n, \ldots$
The very famous Fibonacci Sequence is defined recursively by
   $a_1 = 1, a_2 = 1$ and $a_n = a_{n-1} + a_{n-2}$:   1, 1, 2, 3, 5, 8, 13, 21, 34, …

An arithmetic sequence is a sequence each term of which, starting from the second, is equal to the preceding term added to a fixed number. An arithmetic sequence can be written in the form a, a + d, a + 2d, a + 3d, … where d is the difference. Note that each term of an arithmetic sequence is equal to the arithmetic mean of its

neighbors: $a_n = \dfrac{a_{n-1} + a_{n+1}}{2}$

The nth term of an arithmetic sequence is given by
   $$a_n = a_1 + d(n - 1) \tag{8}$$

The sum of n terms of an arithmetic sequence is given by

$$S_n = \frac{(a_1 + a_n)n}{2} \quad \text{or the equivalent formula} \quad S_n = \frac{(2a_1 + d(n - 1))n}{2} \tag{9}$$

2

A geometric sequence is a sequence each term of which, starting from the second, is equal to the preceding term multiplied by a fixed number.

A geometric sequence can be written in the form b_1, b_1r, b_1r^2, b_1r^3,... where r is a ratio. Note that each term of a geometric sequence, starting from the second, is a geometric mean of its neighbors: $b_n = \sqrt{b_{n-1} \cdot b_{n+1}}$.

The nth term of a geometric sequence is given by

$$b_n = b_1 r^{n-1} \qquad (10)$$

and the sum of n terms is given by the formula

$$S_n = \frac{b_1(r^n - 1)}{r - 1}. \qquad (11)$$

Note that if a sequence is infinite and decreasing, i.e. $|r| < 1$, then the formula for the sum of all terms is

$$S = \frac{b_1}{1 - r}. \qquad (12)$$

IV. CLOCK ARITHMETIC AND COMPARISON BY MODULO

Imagine that the number line is wrapped around a wheel and the full revolution is divided into 12 equal parts.

Now we have constructed a circular number line to make operations with numbers in 12-clock arithmetic.

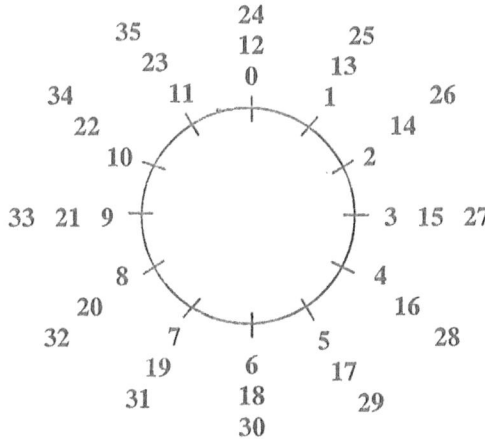

Let us agree that the numbers on this scale represented by the same point on the circle are congruent: $1 \cong 13$, $2 \cong 14$, $11 \cong 23$, etc.

Notice that all congruent numbers have the same remainder if we divide the number by 12.

The universally adopted notation for congruent numbers is:

$1 \cong 13 \pmod{12}$, $2 \cong 14 \pmod{12}$, $11 \cong 23 \pmod{12}$.

In general $a \cong b \pmod{m}$ means that $a - b$ is divisible by m or, in other words, the division of a and b by m gives the same remainder.

For example in 5-clock arithmetic (a circle would be divided into 5 equal parts), $0 \cong 5 \pmod 5$, $1 \cong 6 \pmod 5$, $7 \cong 12 \pmod 5$, $34 \cong 24 \pmod 5$, and so on.

Congruence and equality of numbers have similar properties. Let us list some of them.

1. Transitive property: If $a \cong b \pmod m$ and $b \cong c \pmod m$, then $a \cong c \pmod m$.

2. If $a \cong b \pmod m$ and $c \cong d \pmod m$, then $a + c \cong b + d \pmod m$.
   This property allows us to add the left and right sides of comparisons of congruent numbers with the same modulo.

3. If $a + c \cong b \pmod m$, then $a \cong b - c \pmod m$.
   This property allows us to remove an addent from one side of a comparision to the other side with the opposite sign.

4. If $a \cong b \pmod m$ and $c \cong d \pmod m$, then $ac \cong bd \pmod m$.
   This property allows us to multiply both parts of a comparison by congruent numbers with the same modulo.

5. As a corollary of the previous property we have:

   If $a \cong b \pmod m$, then $a^n \cong b^n \pmod m$.

Let us illustrate by examples how to use these properties to solve some problems.

a) Find the remainder after dividing 5^{20} by 24.
   Solution:
   $5^2 \cong 1 \pmod{24}$. By property 4 : $(5^2)^{10} \cong 1^{10} \pmod{24} \Rightarrow 5^{20} \cong 1 \pmod{24}$.
   Thus the remainder after dividing 5^{20} by 24 is 1.

b) Prove that $2^{60} + 7^{30}$ is divisible by 13.
   Solution:
   $2^{60} = (2^4)^{15}$; $2^4 = 16 = 13 + 3 \Rightarrow 2^4 \cong 3 \pmod{13} \Rightarrow 2^{60} \cong 3^{15} \pmod{13}$.
   $7^{30} = (7^2)^{15}$; $7^2 = 49 = 13 \cdot 4 - 3 \Rightarrow 7^2 \cong -3 \pmod{13} \Rightarrow 7^{30} \cong -3^{15} \pmod{13}$.
   By property 2 , $2^{60} + 7^{30} \cong 3^{15} - 3^{15} \pmod{13} \cong 0 \pmod{13}$.
   Thus $2^{60} + 7^{30}$ is divisible by 13.

4

PROBLEMS

1. Prove that $p^2 - 1$, where p is prime and $p \geq 5$, is divisible by 24.

2. How many zeroes are at the end of the number $1 \cdot 2 \cdot 3 \cdot 4 \cdot 5 \cdot \ldots \cdot 99 \cdot 100$?

3. Find the smallest number such that the quotient after division by 2 is a perfect square, and the quotient after division by 3 is a perfect cube.

4. Find a 4-digit number such that if you read this number from right to left, the new number is four times larger than the original number.

5. Find a two-digit number that is equal to the sum of its tens digit and the square of its units digit.

6. Find a two-digit number that is equal to twice the product of its digits.

7. Find the digits in the next operation. (The same letters denote the same digits and different letters denote different digits.)

   a)
   ```
     ABCD
   + EFGB
   ------
    EFCBH
   ```

   b)
   ```
     SEND
   + MORE
   ------
    MONEY
   ```

8. Find the missing digits.

   a)
   ```
      *3*
       *7
     ----
      ***
     *0**
    ------
    ****8
   ```

   b)
   ```
     *1***
       237
    -------
    ******
     *****
     *****
    -------
    7***065
   ```

   c)
   ```
                 ____*_*_8_*_*__
     ** |*******
         ***
        ----
          **
          **
        ----
         ***
         ***
        ----
   ```

9. Prove that $11^{10} - 1$ is divisible by 100.

10. Prove that $2^{2002} - 1$ is divisible by 3.

11. Prove that for any odd natural number n the number $n^{12} - n^8 - n^4 + 1$ is divisible by 512.

12. Prove that a number $n^5 - 20n^4 + 64n$, where n is an even number larger than 4, is divisible by 1920.

13. Prove that for any natural numbers m and n, one of the three numbers m − n, m + n and mn is divisible by 3.

14. Find a two-digit number that is equal to the sum of the square of its tens digit and the cube of its units digit.

15. Prove that for an even n, a number $n^5 - 5n^3 + 4n$ is divisible by 1440.

16. Prove that $43^{17} + 17^{17}$ is divisible by 60.

17. Prove that for even n, $\dfrac{n}{12} + \dfrac{n^2}{8} + \dfrac{n^3}{24}$ is a natural number.

18. A father brought some apples from the garden. His children asked him how many apples he brought. The father answered that he did not count them but noticed that, when he tried to take them 2 at a time, 3 at a time, 4 at a time, 5 at a time, 6 at a time, or 7 at a time, every time 1 apple remained. What was the minimum possible number of apples the father brought from the garden?

19. Prove that the difference between a three-digit number and the same number written from right to left cannot be the square of a natural number.

20. A dog chases a fox. At the same time the dog makes 2 jumps, the fox makes 3 jumps, but the jump of the fox is equal to 1 meter and the jump of the dog is twice as long. What distance does the dog have to cover to catch the fox, if the distance between them at the beginning is 50 meters?

21. Find the value of $1^2 - 2^2 + 3^2 - 4^2 + 5^2 - 6^2 + \ldots + 999^2 - 1000^2$.

22. The sum of three consecutive terms of the arithmetic sequence is 5 times less than their product. Find these terms if they are integers.

23. Evaluate $\sqrt{5\sqrt{3\sqrt{5\sqrt{3\sqrt{5\sqrt{3}\ldots}}}}}$

24. In an arithmetic sequence the sum of the $6^{th}, 9^{th}, 12^{th}$ and 15th terms is equal to 20. Find the sum of the first 20 terms.

25. The sum of the terms of an infinite decreasing geometric sequence is equal to 15. The sum of the squares of its terms is 75. Find the sum of the first 6 terms of this sequence.

26. Find the sum $1 + 11 + 111 + 1111 + \ldots + \underbrace{111\ldots1}_{n}$.

6

27. Find the sum of the following sequence if $|x|<1$ and evaluate the sum at $x = 3/5$

$$1 + 3x + 5x^2 + 7x^3 + 9x^4 + \dots + (2n\text{-}1)x^{n\text{-}1} + \dots$$

28. Not performing the indicated operations and without using a calculator, determine whether the following fraction is proper or improper:

$$\frac{244 \cdot 395 - 151}{244 + 395 \cdot 243}$$

29. The sum of a two-digit number and the number written with the same digits from right to left is equal to a perfect square. Find all such numbers.

30. Assuming that the hands of a clock are moving with constant speed, find how long it takes for the minute hand to catch up to the hour hand after the clock shows precisely 3:00.

31. Prove that the remainder after dividing the square of an odd number by 8 is 1.

32. Prove that the expression $2a^2 + 2b^2$ is equal to the sum of two squares.

33. Show that the product of three consecutive whole numbers added to the second of them is equal to the cube of the second number.

34. Evaluate the expression $\dfrac{a - b}{\sqrt{b}}x^2 - 2ax + a\sqrt{b}$, at $x = \dfrac{\sqrt{ab}}{\sqrt{a - b}}$.

35. Prove that $\sqrt{2+\sqrt{3}}\sqrt{2+\sqrt{2+\sqrt{3}}}\sqrt{2+\sqrt{2+\sqrt{2+\sqrt{3}}}}\sqrt{2-\sqrt{2+\sqrt{2+\sqrt{3}}}} = 1.$

36. Evaluate without using a calculator:

$$(\sqrt[6]{9 + 4\sqrt{5}} + \sqrt[3]{2 + \sqrt{5}}) \cdot \sqrt[3]{2 - \sqrt{5}}.$$

37. Prove that if $a + b + c = 0$, then $a^3 + b^3 + c^3 = 3abc$.

38. Prove that $1 + 3^x + 9^x$ is divisible by 13 if $x = 3k + 1$, where k is a natural number.

39. Prove that the sum of the cubes of three consecutive whole numbers is divisible by 9.

40. Prove that $n^3 + 11n$ is divisible by 6 for any integer n.

41. Find $3^x - 3^{-x}$ if $9^x + 9^{-x} = 51$.

42. Prove that if from a three-digit number you subtract the number written with the same digits in the opposite direction, then the sum of the digits of the difference is equal to 18.

43. There are 80 of the same kind of coins on a table. One of them is false and lighter than the others. How can you find this coin using a balanced scale (no weights) and making only four weighings.

44. Is it possible to make a right triangle from 36 matches, without breaking any of them?

45. Prove that the number $5^{5k+1} + 4^{5k+2} + 3^{5k}$ is divisible by 11 for any natural number k.

46. The number $3^{105} + 4^{105}$ is divisible by 13, 49, 181 and 379, and is not divisible either by 5 or by 11. How can this result be confirmed?

47. A total of 6869 digits were used to number the pages of an encyclopedic dictionary. How many pages does this dictionary contain?

48. Find the sum $\dfrac{1}{1 \cdot 2} + \dfrac{1}{2 \cdot 3} + \dfrac{1}{3 \cdot 4} + \dfrac{1}{4 \cdot 5} + ... + \dfrac{1}{99 \cdot 100}$.

49. Prove that $\dfrac{1}{\log_2 \pi} + \dfrac{1}{\log_5 \pi} > 2$ without using a calculator.

50. Is it possible to write the number 203 as a sum of natural numbers, and simultaneously as a product of the same numbers?

51. The sequence given by $a_1 = 7$ and a_{n+1} is the sum of the digits in a_n^2. Find a_{1000}.

52. Without using a calculator, decide which is larger:
    a) 3^{500} or 7^{300} b) $\log_5 6$ or $\log_6 7$

53. Nine identical pens cost $11 and some cents. Thirteen of the same pens cost $15 and some cents. What is the price of one pen?

54. Prove that the following numbers are composite:
    a) $2^{3^{1987}} + 1$;   b) $2^{3^{1987}} - 1$.

55. Find three prime numbers such that their sum is 5 times less than their product.

56. Suppose that the lengths of the sides of a right triangle are natural numbers. Is it possible that the lengths of the legs are both odd numbers?

57. Prove that the expression $\dfrac{10^n + 8}{9}$ is a whole number for any $n \in \mathbb{N}$.

58. Simplify $a^{\log_c b} - b^{\log_c a}$.

59. Prove that $\log_2 5 = \log_2 3 \cdot \log_3 4 \cdot \log_4 5$.

60. Find any three consecutive numbers such that each of them is divisible by the square of a natural number larger than 1.

61. Factor the following trinomials:

   a) $x^8 + x^4 + 1$ into 3 factors
   b) $x^5 + x + 1$ into 2 factors
   with integer coefficients.

62. Factor $x^9 + x^4 - x - 1$ into 5 factors with integer coefficients.

63. If p is a prime number, then $8p^2 + 1$ is a prime number only for $p = 3$. Prove.

64. The sum of two numbers is 15 and their arithmetic mean is 25% more than their geometric mean. Find the numbers.

65. Rationalize the denominator of $\dfrac{1}{1 + \sqrt{2} + \sqrt{3}}$.

66. Evaluate $x^3 + 3x - 14$ at $x = \sqrt[3]{7 + 5\sqrt{2}} - \dfrac{1}{\sqrt[3]{7 + 5\sqrt{2}}}$.

67. Compare two numbers a and b (without a calculator) if
   $a = \dfrac{9}{\sqrt{11} - \sqrt{2}}$, $b = \dfrac{6}{3 - \sqrt{3}}$.

68. Prove that $\sqrt{\underbrace{111...1}_{2n} - \underbrace{222...2}_{n}} = \underbrace{333...3}_{n}$.

69. Prove that $x^2 + 3x + 5$ is not divisible by 121 for any integer x.

Solve each equation for x.

70. $(x^2 - x + 1)^4 - 10x^2(x^2 - x + 1)^2 + 9x^4 = 0$

71. $(\sqrt{2 + \sqrt{3}})^x + (\sqrt{2 - \sqrt{3}})^x = 4$

72. $\sqrt{a - \sqrt{a + x}} = x$

73. $x^2 + 2\sqrt{x^2 - 3x + 11} = 3x + 4$

74. $\sqrt[3]{x^2 \sqrt[3]{x^2 \sqrt[3]{x^2 \ldots}}} = 49$

75. At what value of a is the inequality $ax^2 + 2ax + 0.5 \geq 0$ valid for all $x \epsilon R$?

76. Solve the following inequality:

$$\sqrt{x^2 - 4x} > x - 3$$

Solve each logarithmic equation.

77. $9^{\log(x-a) - \log 2} = 3^{\log(x-1)}$

78. $\dfrac{\log x}{\log(3 - 4x)} = 2$

79. $\log_2(9^{x-1} + 7) = 2\log_2(3^{x-1} + 1)$

80. $\log^2 x - \log x^2 + 1 = 0$

81. $\log_4^2 x + \log_4 \sqrt{x} - 1.5 = 0$

82. $\dfrac{1}{\log x + 1} + \dfrac{6}{\log x + 5} = 1$

83. $\log_3 x - 2\log_{1/3} x = 6$

84. $\log_2(9 - 2^x) = 3 - x$

85. $\log_2(25^{x+3} - 1) = 2 + \log_2(5^{x+3} + 1)$

86. $\log_2(4^x + 4) = \log_2 2^x + \log_2(2^{x+1} - 3)$

Solve each exponential equation.

87. $\left(\dfrac{1}{7}\right)^{2x^2 + x - 0.5} = \dfrac{\sqrt{7}}{7}$

88. $2 \cdot 3^{x+1} - 3^x = 15$

89. $5^{x+1} = 8^{x+1}$

90. $3^x + 3^{3-x} = 12$

91. $4^{\sqrt{x-2}} + 16 = 10 \cdot 2^{\sqrt{x-2}}$

92. $9^x - 8 \cdot 3^x - 9 = 0$

Solve each system of equations.

93. $\begin{cases} \log_{1/3}(x + y) = 2 \\ \log_3(x - y) = 2 \end{cases}$

94. $\begin{cases} \log(x^2 + y^2) = 1 + \log 13 \\ \log(x + y) = \log(x - y) + \log 8 \end{cases}$

95. $\begin{cases} 3^y \cdot 9^x = 81 \\ \log(x+y)^2 - \log x = 2\log 3 \end{cases}$

96. $\begin{cases} 10^{1+\log(x+y)} = 50 \\ \log(x+y) + \log(x-y) = 2 - \log 5 \end{cases}$

97. $\begin{cases} 3^x \cdot 2^y = 576 \\ \log_{\sqrt{2}}(y-x) = 4 \end{cases}$

Solve each inequality.

98. $4^{5-2x} \le 0.25$

99. $0.4^{2x+1} > 0.16$

100. $3^{4x+3} \le (\frac{1}{9})^{x/2}$

101. $4^x - 2^{x+1} - 8 > 0$

Solve each system of equations.

102. $\begin{cases} xy = 6 \\ yz = 12 \\ zx = 8 \end{cases}$

103. $\begin{cases} xy + yz = 8 \\ yz + zx = 20 \\ zx + xy = 14 \end{cases}$

104. $\begin{cases} x^2 y - xy^2 = 30 \\ xy^2 + x^2 y = 70 \end{cases}$

105. $\begin{cases} x^4 + y^4 = 17 \\ x + y = 3 \end{cases}$

107. $\begin{cases} x^2 + xy + y^2 = 7 \\ x + xy + y = -5 \end{cases}$

107. $\begin{cases} x - y + \sqrt{x^2 - 4y^2} = 2 \\ x^5 \sqrt{x^2 - 4y^2} = 0 \end{cases}$

108. $\begin{cases} x^2 = (x - a)\, y \\ y^2 - xy = 9ax \end{cases}$

109. $\begin{cases} 2(x + y) = 5xy \\ 8(x^3 + y^3) = 65xy \end{cases}$

110. $\begin{cases} x^4 + y^4 - x^2 y^2 = 13 \\ x^2 - y^2 + 2xy = 1 \end{cases}$

111. At what value of k does the system of equations

$\begin{cases} (k - 2)x + 27y = 4.5 \\ 2x + (k + 1)y = -1 \end{cases}$

have infinitely many solutions?

112 A father is now 5 times older than his son. The father graduated from the university at the age of 22. The time after this event is equal to half of the time for his son to reach the age of 22. How old are the father and the son now?

113. Two bicyclists ride along a circle with a circumference of 800 meters in the same and then in opposite directions. In the first case one of them catches another every 24 minutes, and in the second case they meet each other every 3 minutes. Find the speed of each bicyclist.

114. A team of workers has to make 360 parts. They made 4 parts more each day than was planned and finished the job 1 day earlier than was expected. How many days did it take the workers to finish the job?

115. Two students are moving downstairs by escalator. One of them counted 40 steps. Another, who moved twice as slow, counted 60 steps. How many steps would they count if the escalator stopped?

116. There are 20 participants in a vacation trip. The common weight of their backpacks is 20 kilograms. The backpack of each adult male weighs 2kg, the backpack of each adult female weighs 0.5kg, and each child has a backpack of 0.25kg. How many men, women and children participated in this trip?

117. For the equation $\sqrt{1995} \cdot x^{\log_{1995} x} = x^2$, find the last three digits of the product of the two roots of this equation without using a calculator.

118. Two friends live in villages A and B. One of them leaves village A at 8:38AM and can arrive at village B at 1:08 PM. The other leaves village B at 10:08 AM and can arrive at village A at 1:44 PM. If both friends leave at their respective times, when will they meet?

119. Two cars simultaneously leave cities A and B and move toward each other. Once they meet car A needs 2 more hours to reach its destination, and car B needs 9/8 hours. Find the speed of each car if the distance between A and B is 210 miles.

120. Find the whole number solutions of the equation $xy = x + y$.

121. Solve the system of equations

$$\begin{cases} x^2 + y^2 - xy = 61 \\ x + y - \sqrt{xy} = 7. \end{cases}$$

122. Find two numbers having these conditions: their sum is 1244; the numbers would be equal if the digit 3 were added at the end of the first number, and the digit 2 erased from the end of the second number.

123. For what value of k is the difference of the roots of the equation

$$2x^2 - (k+1)x + k + 3 = 0 \quad \text{equal to 1?}$$

124. For the equation $(k^2 - 5k + 3)x^2 + (3k - 1)x + 2 = 0$, find a value k so that one root of the equation is twice the other.

125. The last digit of a three-digit number is 3. If you move this digit to the front of the number, the new number will be 1 larger than triple the original number. Find the original number.

126. Solve the equation $(x + 1)(x + 2)(x + 3)(x + 4) = 24$.

127. Solve the equation $(x + a)(x + 2a)(x + 3a)(x + 4a) = b^4$.

128. Two couriers leave cities A and B simultaneously walking toward each other. They meet at a point that is 12 miles from city B. Then they continue to walk to the opposite city, and after reaching their point of destination immediately turn back. The second time they meet at a point which is 6 miles from city A in 6 hours after their first meeting. Find the distance between A a B and the speed of each courier.

129. A pool has 4 taps. If all four of them are turned on, then the pool can be filled in 4 hours. The first, second and third taps working together can fill the pool in 5 hours. The second, third and fourth together can do it in 6 hours. How long does it take to fill the pool by using only the first and fourth taps?

130. Prove that if $\dfrac{x^2+y^2}{xy}+\dfrac{4xy}{x^2+y^2}=4$, then $x=y$.

131. In the equation $(x^2+\ldots)(x+1)=(x^4+1)(x+2)$ one number is erased (shown by $\ldots$). Find this number if it is known that one of the roots of this equation is 1.

132. Find 7 integer solutions of the equation $y^2=6(x^3-x)$.

133. Solve the system of equations

$$\begin{cases} 2y=4-x^2 \\ 2x=4-y^2. \end{cases}$$

134. Find all solutions of the system of equations

$$\begin{cases} x+y=2 \\ xy-z^2=1. \end{cases}$$

135. During spring an athlete lost 25% of his weight, and during summer he gained 20%. During fall he lost 10%, and during winter he gained 20%. Did he have an overall gain or loss in his weight for this year?

136. A student has to solve 20 problems. For each correctly solved problem he earns 8 points. For each incorrect problem, he loses 5 points. For each problem he did not even try to solve, he gets 0 points. The total of his points is 13. How many problems did the student try to solve?

137. A flock of white geese flew over a chain of lakes. On each lake half of all geese and half of a goose set down. The rest of the geese continued to fly farther. All geese set down on the 7 lakes. How many geese were in the flock?

138. Write a quadratic equation with integer coefficients if one known root of this equation is $\dfrac{\sqrt{3}-\sqrt{5}}{\sqrt{3}+\sqrt{5}}$.

139. A boat covers 10 miles with the current of a river and then 6 miles against the current. The speed of the current is 1 mile per hour. It is planned that the entire trip will take from 3 to 4 hours. In what interval should the speed of the boat be without the current, to make this trip as planned?

140. Two workers take 12 days to do a job. If the first worker does half of the job and the second the other half of the job, they will finish the job in 25 days. How long would it take for each worker to do this job alone?

GEOMETRY

Three different kinds of problems in Geometry are included in this section: calculating problems, proving problems and construction problems. Let us recall that for constructions you can use only a straight edge and compass.

Let us also list popular formulas useful for solving many problems. If you don't know some of these, try to prove them as additional "proof" problems.

I. Areas of Popular Figures.

1. Area of a Triangle

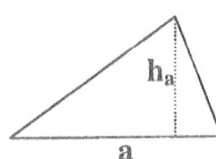 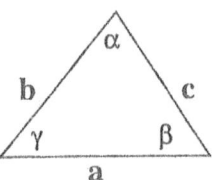

$$S = \frac{1}{2} ah_a \qquad (13)$$

$$S = \frac{1}{2} ab\sin\gamma \qquad (14)$$

$$S = \sqrt{p(p-a)(p-b)(p-c)} \quad \text{where} \quad p = \frac{a+b+c}{2} \quad \text{(Heron's Formula)} \qquad (15)$$

$$S = pr \text{ where } r \text{ is a radius of the inscribed circle} \qquad (16)$$

$$S = \frac{abc}{4R} \text{ where } R \text{ is a radius of the circumscribed circle} \qquad (17)$$

2. Area of a Quadrilateral

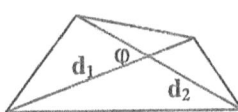 $$S = \frac{1}{2} d_1 d_2 \sin\varphi \qquad (18)$$

3. Area of a Parallelogram

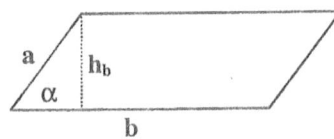 $$S = h_b \cdot b \qquad (19)$$

$$S = ab\sin\alpha \qquad (20)$$

4. Area of a Trapezoid

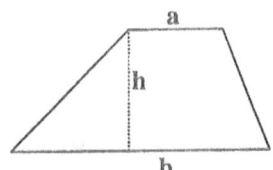 $$S = \frac{1}{2} h(a+b) \qquad (21)$$

II. Properties of the Elements in a Triangle.

1. Medians

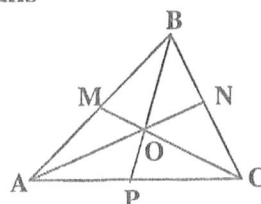

$$\frac{AO}{ON} = \frac{BO}{OP} = \frac{CO}{OM} = \frac{2}{1} \qquad (22)$$

2. Bisectors

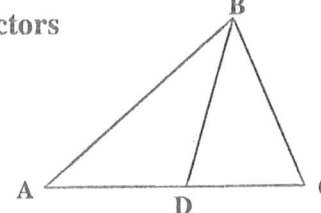

$$\frac{AB}{BC} = \frac{AD}{DC} \qquad (23)$$

3. Altitudes

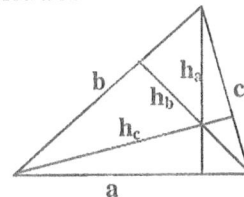

$$h_a : h_b : h_c = \frac{1}{a} : \frac{1}{b} : \frac{1}{c} \qquad (24)$$

III. Solving Triangles.

1. Right Triangle Trigonometry

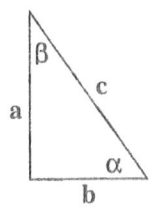

$$a = c\sin\alpha$$
$$b = c\cos\alpha \qquad (25)$$
$$a = b\tan\alpha$$

2. Sine Theorem

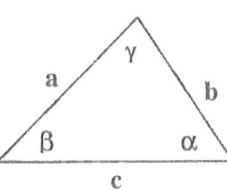

$$\frac{a}{\sin\alpha} = \frac{b}{\sin\beta} = \frac{c}{\sin\gamma} = 2R \qquad (26)$$

where R is a radius of the circumscribed circle.

3. Cosine Theorem

$$c^2 = a^2 + b^2 - 2ab\cos\gamma \qquad (27)$$

IV. Quadrilaterals and Circles

1. Circumscribed Circle

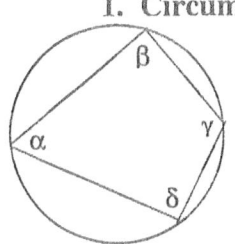

$$\alpha + \gamma = \beta + \delta = 180° \qquad (28)$$

2. Inscribed Circle

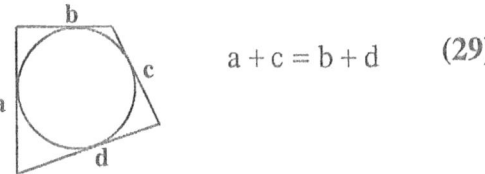

$$a + c = b + d \qquad (29)$$

PROBLEMS

141. Prove that the bisectors of all external angles of a rectangle form a square.

142. A segment EF through the point of intersection of the diagonals of trapezoid ABCD is parallel to the bases of the trapezoid. Prove that EP = PF.

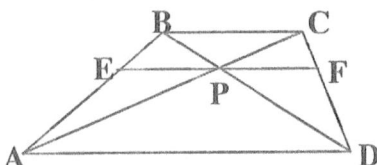

143. The diagonals of 12cm and 5cm in trapezoid ABCD are perpendicular. Find the length of the midsegment of this trapezoid.

144. Fifteen billiard balls are located on a plane in such a way that three tangent lines to them form an equilateral triangle. Find the side of this triangle if the diameter of each ball is 5cm.

145. In a right triangle ABC with $\angle B = 90°$, the altitude from the vertex of the right angle is a and the median from the same vertex is b. Find the area of this triangle.

146. A circle is circumscribed about a square with side a. The circle is inscribed in a regular hexagon. Find the area of this hexagon.

147. Find the area of the shaded part of a semicircle of radius R.

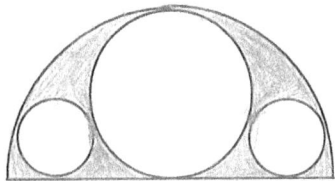

148. Through a given point inside a circle draw the chord of minimal length.

149. BK is a median of right triangle ABC, where $\angle B$ is right and $\angle C = 30°$. Find the area of $\triangle BCK$ if AB = 4in.

150. In a rhombus ABCD, a side is a geometric mean of its diagonals. Find the angles of this rhombus.

151. In a triangle with sides 11in, 12in, and 13in, find the length of the median to the largest side.

152. Find the area of an isosceles trapezoid if its diagonal has length l, and the angle between the diagonal and larger base is α.

153. A square is inscribed in a right triangle with legs a and b, in such a way that the square and the triangle share the right angle. Find the perimeter of the square.

154. In triangle ABC, AB = BC = 4in. The median to BC is 3in. Find the base of the triangle.

155. Find the angle between two medians from the vertices of the acute angles of an isosceles right triangle.

156. The distance between the points of tangency of two tangent lines to a circle from a point outside the circle is 14.4cm. The length of a tangent segment is 12cm. Find the radius of the circle.

157. What are the legs of a right triangle with hypotenuse c, in order to obtain the triangle of maximum area?

158. A parallelogram of maximum area is inscribed in an isosceles triangle with sides of 15in, 15in, and 18in, in a such way that the parallelogram and triangle share the same angle at the base. Find the sides of the parallelogram.

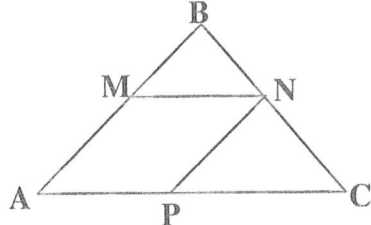

159. Find the area of a circle inscribed in an isosceles trapezoid with given bases a and b. Then find the diagonal of this trapezoid.

160. A square is inscribed in a semicircle of radius R. Find the side of this square.

161. Prove that if the sides of a right triangle are the terms of an arithmetic sequence, then the difference of this sequence is equal to the radius of a a circle inscribed in this triangle.

162. A circle is inscribed in an isosceles trapezoid with the area of $2in^2$. Find the sides of this trapezoid if an angle at the base is equal to $30°$.

163. A circle is inscribed in an isosceles trapezoid. The ratio of the distance between the intersection point of the diagonals and the center of the circle to the radius of the circle is the same as 3:5. Find the ratio of the perimeter of the trapezoid to the circumference of the circle.

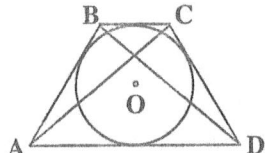

164. The sides of a triangle with an angle of $120°$ are the terms of an arithmetic sequence whose difference is equal to 1. Find the sides of the triangle.

165. Among all isosceles trapezoids with an acute angle of $45°$ whose sum of the larger base and an altitude is a, find the trapezoid of largest area.

166. Prove that the sum of the distances from an arbitrary point on a circle to the two neighboring vertices of the inscribed equilateral triangle is equal to the distance from this point to the third vertex of the triangle.

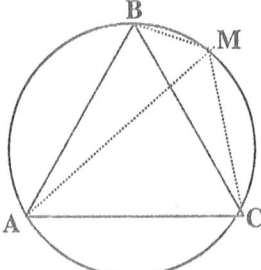

Given: $AB = BC = AC$ Prove: $BM + MC = MA$

167. An altitude, a bisector, and a median from the same vertex of a triangle divide the angle at this vertex into four equal parts. Prove that this is a right triangle.

168. Prove that if in a triangle an altitude and a median divide an angle into three equal parts, then the triangle is a right triangle.

169. A trapezoid is inscribed in a semicircle of radius R. What is the measure of the angle at the base of the trapezoid of maximal perimeter?

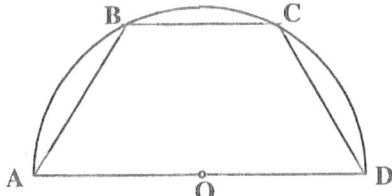

170. The diagonal divides an angle of a parallelogram into two angles of 84° and 36°. What are the measures of the angles formed by the other diagonal at the vertex of the parallelogram?

171. Three equal squares are joined in a row and the left bottom vertex is connected with the right top vertices of the second and third squares. Prove that the sum of the angles between these segments and the top base of the squares is equal to 45°.

Prove: $\alpha + \beta = 45°$

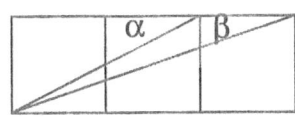

172. Construct a right triangle given the measure of an acute angle and the length of the opposite leg.

173. Construct an isosceles triangle given the lengths of an altitude to the base and a median to one of the congruent sides.

174. Construct a circle of given radius R which intersects a given straight line and forms a chord of given length a.

175. Construct a quadrilateral given the lengths of its three sides and the measures of the two angles containing the fourth side.

176. Construct a trapezoid given the lengths of its four sides.

177. Construct a circle inscribed in the given angle which passes through the given point.

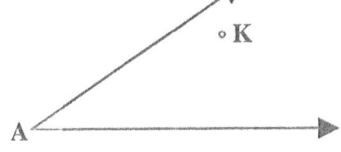

178. Three points are marked on a plane. Two of them are the midpoints of the sides of a triangle. The third point is the intersection point of the medians of this triangle. The rest of the elements of the triangle are erased. Restore the triangle.

179. Find the angles of a trapezoid ABCD whose larger base is a diameter of a semicircle, whose nonparallel sides intersect the semicircle at midpoints, and whose smaller base is a tangent line to this semicircle.

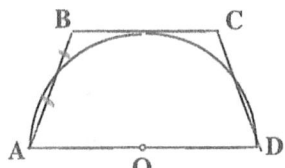

180. A section of a cube passes the center of the cube and two midpoints of the edges from a common vertex. Find the area of the section if the edge of the cube is a.

181. A cone is inscribed in a sphere. The slant segment of the cone is equal to the diameter of its base. Find the ratio of the surface area of the cone to the surface area of the sphere.

182. The base of a right prism is an isosceles right triangle with leg a. The angle between the diagonal of the largest lateral face and the diagonal of the other lateral face, originating from the same vertex, is α. Find the volume of the prism.

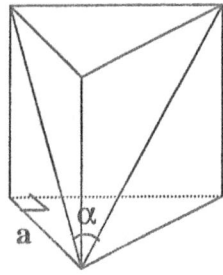

183. The measure of an angle between a diagonal of a regular quadrilateral prism and a lateral face is 30°. Find the angle between the diagonal and the base of the prism.

184. A section of a regular quadrilateral prism with a side of 2cm and a height of 4cm contains a diagonal of the base, and is parallel to the diagonal of the prism. Find the area of the section.

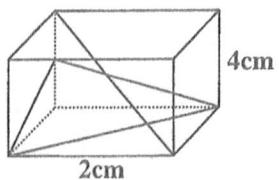

185. The distances from point M inside an angle of 60° to the sides of the angle are 2cm and 11cm. Find the distance from point M to the vertex of the angle.

186. If the diagonals of a trapezoid are perpendicular, then the sum of the squares of the diagonals is equal to the square of the sum of the bases. Prove.

187. If the midpoints of the sides of two convex quadrilaterals match, then the areas of these quadrilaterals are equal. Prove.

188. What angles does an isosceles trapezoid have, if its diagonal cuts the trapezoid into two isosceles triangles?

189. In trapezoid ABCD with the intersection point of the diagonals O, the areas of triangles AOD and BOC are equal to S_1 and S_2, respectively. Find the area of the trapezoid.

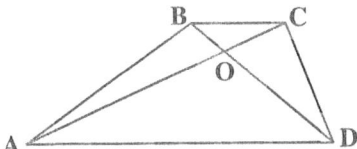

190. In a right triangle the product of the lengths of all altitudes is 4 times less than the product of the lengths of its sides. Find the angles of the triangle.

191. In a right triangle with legs a and b, find the bisector of the triangle from the vertex of the right angle.

192. The hypotenuse of a right triangle with legs a and b is a side of the square constructed outside of the triangle. Find the distance from the center of the square to the vertex of the right angle of the triangle.

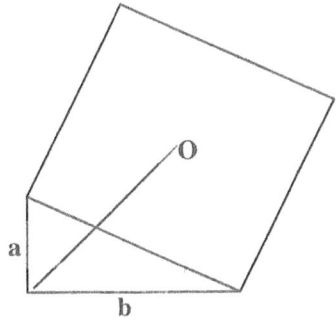

193. Is it possible to have two triangles such that the length of each side of one triangle is less than 1 meter, the length of each side of the other triangle is more than 100 meters, and the area of the first triangle is larger than the area of the second?

194. Prove that in any triangle with sides a, b, and c, the length of the median to the side b can be found by the formula $m_b = \frac{1}{2}\sqrt{2a^2 + 2c^2 - b^2}$.

TRIGONOMETRY

Let us supply a collection of formulas and tables which are valuable to prove trigonometric identities and solve trigonometric equations. There are many formulas in Trigonometry; very often students are scared to death because they believe it is impossible to remember these formulas. This is not true. If we explain the rules of memorizing and give a minimum of useful formulas, with understanding the derivations and the goals of using these formulas, then everyone can memorize them without too much effort. Sometimes it is very useful to look at the set of formulas all together instead of separately. It is very important, of course, to understand when to use the formulas.

I. Formulas For Trigonometric Functions In All Quadrants In Terms Of Those In Quadrant I

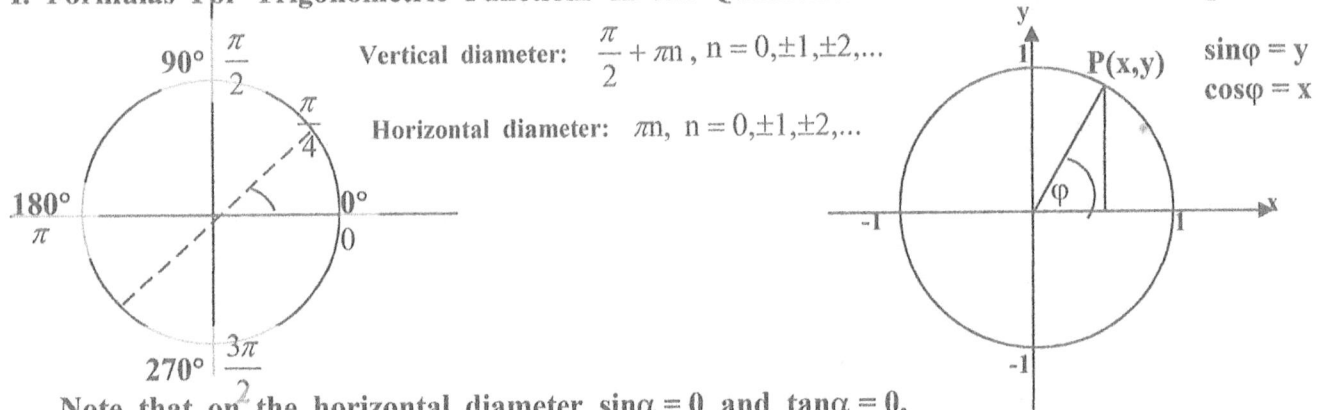

Vertical diameter: $\dfrac{\pi}{2} + \pi n$, $n = 0, \pm 1, \pm 2, \ldots$

Horizontal diameter: πn, $n = 0, \pm 1, \pm 2, \ldots$

$\sin\varphi = y$
$\cos\varphi = x$

Note that on the horizontal diameter $\sin\alpha = 0$ and $\tan\alpha = 0$, and on the vertical diameter $\cos\alpha = 0$ and $\cot\alpha = 0$.

Rule for memorizing formulas :

Let $0° < \alpha < 90°$, i.e. α is an angle in quadrant I.

1. If an angle is added (or subtracted) to an angle on the horizontal diameter, then the name of the function in the answer is the same, and the sign of the answer is the sign of the original function in the corresponding quadrant.

   Examples: $\sin(180°-\alpha) = \sin\alpha$ since α is subtracted from an angle on the horizontal diameter and $(180°-\alpha)$ is an angle in quadrant II, where the sine is positive. $\tan(360°-\alpha) = -\tan\alpha$ (please, explain why).

2. If an angle is added (or subtracted) to an angle on the vertical diameter, then the name of the function has to be changed to the name of the cofunction, and the sign of the answer is the sign of the original function in the corresponding quadrant.

   Examples: $\cos(90°+\alpha) = -\sin\alpha$ since α is added to an angle on the vertical diameter and $(90°+\alpha)$ is an angle in quadrant II, where the cosine is negative. $\cot(270°-\alpha) = \tan\alpha$ (explain why).

   Notice: In the formulas we suppose that an angle α is acute, but the formulas are valid for any α.

24

II. Table of Exact Values of Trigonometric Functions of Special Angles

	30°(π/6)	45°(π/4)	60°(π/3)
sin	$\dfrac{1}{2}$	$\dfrac{\sqrt{2}}{2}$	$\dfrac{\sqrt{3}}{2}$
cos	$\dfrac{\sqrt{3}}{2}$	$\dfrac{\sqrt{2}}{2}$	$\dfrac{1}{2}$
tan	$\dfrac{1}{\sqrt{3}}$	1	$\sqrt{3}$

(30)

II. ANGLE ADDITION FORMULAS

(31)

$$\sin(\alpha \pm \beta) = \sin\alpha \cdot \cos\beta \pm \sin\beta \cdot \cos\alpha$$

$$\cos(\alpha \pm \beta) = \cos\alpha \cdot \cos\beta \mp \sin\alpha \cdot \sin\beta$$

$$\tan(\alpha \pm \beta) = \frac{\tan\alpha \pm \tan\beta}{1 - \tan\alpha \cdot \tan\beta}$$

Corollary 1: DOUBLE ANGLE FORMULAS

$$\sin 2\alpha = 2\sin\alpha \cdot \cos\alpha$$

(32)

$$\cos 2\alpha = \cos^2\alpha - \sin^2\alpha$$

$$\tan 2\alpha = \frac{2\tan\alpha}{1 - \tan^2\alpha}$$

Corollary 2: HALF ANGLE FORMULAS

$$\sin\frac{\alpha}{2} = \pm\sqrt{\frac{1 - \cos\alpha}{2}}$$

(33)

$$\cos\frac{\alpha}{2} = \pm\sqrt{\frac{1 + \cos\alpha}{2}}$$

$$\tan\frac{\alpha}{2} = \pm\sqrt{\frac{1 - \cos\alpha}{1 + \cos\alpha}}$$

III. SUM AND DIFFERENCE OF TRIGONOMETRIC FUNCTIONS

$$\sin\alpha + \sin\beta = 2\sin\frac{\alpha+\beta}{2}\cos\frac{\alpha-\beta}{2}$$

$$\sin\alpha - \sin\beta = 2\cos\frac{\alpha+\beta}{2}\sin\frac{\alpha-\beta}{2}$$

$$\cos\alpha + \cos\beta = 2\cos\frac{\alpha+\beta}{2}\cos\frac{\alpha-\beta}{2}$$

$$\cos\alpha - \cos\beta = -2\sin\frac{\alpha+\beta}{2}\sin\frac{\alpha-\beta}{2}$$

IV.   Note that $\cos x$ is an even function, i.e. $\cos(-x) = \cos x$, and
$\sin x$, $\tan x$ are odd functions, i.e. $\sin(-x) = -\sin x$, $\tan(-x) = -\tan x$.
This observation allows us to make the following conclusions:
The graph of the function $y = \cos x$ is symmetric with respect to the y - axis, and the
graphs of the functions $y = \sin x$ and $y = \tan x$ are symmetric with respect to
the origin.
As for the functions $\sec x$, $\csc x$ and $\cot x$, we have to remember only that they are
reciprocal to $\cos x$, $\sin x$ and $\tan x$, respectively.

V. SOLUTIONS OF THE SIMPLEST TRIGONOMETRIC EQUATIONS

1. $\sin x = a$, $|a| \le 1$

(34)

$$x = (-1)^n \sin^{-1}a + \pi n, \quad n = 0, \pm1, \pm2, \pm3,\ldots$$

2. $\cos x = a$, $|a| \le 1$

(35)

$$x = \pm\cos^{-1}a + 2\pi n, \quad n = 0, \pm1, \pm2, \pm3,\ldots$$

3. $\tan x = a$

(36)

$$x = \tan^{-1}a + \pi n, \quad n = 0, \pm1, \pm2, \pm3,\ldots$$

PROBLEMS

195. Evaluate without using a calculator : $\dfrac{\log(2\cos15°)}{\log(2\sin15°)}$

Prove the following identities.

196. $\cos20°\cos40°\cos80° = \dfrac{1}{8}$

197. $16\sin10°\sin30°\sin50°\sin70°\sin90° = 1$

198. $\sin47° + \sin61° - \sin11° - \sin25° = \cos7°$

Solve the following equations.

199. $\cos x = \tan x(1 + \cos2x)$

200. $\cos x - \tan^2\dfrac{x}{2} = -\dfrac{3}{2}$

201. $\sin^4 x - \cos^4 x = \cos(\dfrac{3}{2}\pi - x)$

202. Find the minimum positive solution of the equation

$\cos x \cdot \cos3x = \cos5x \cdot \cos7x$.

203. Solve the equation $\quad \cot x - \tan x = \sin x + \cos x$.

204. Solve the system of equations

$\begin{cases} \tan x + \tan y = 2 \\ 2\cos x \cdot \cos y = 1 \end{cases}$

205. Prove that $\quad \cos x + \sqrt{3}\sin x \le 2$.

206. Evaluate without using a calculator: $\cos\dfrac{\pi}{5} \cdot \cos\dfrac{2\pi}{5}$.

207. Prove that if $\alpha = \arctan\dfrac{1}{7}$ and $\beta = \arctan\dfrac{1}{3}$, then $\cos2\alpha = \sin4\beta$.

27

FUNCTIONS

208. Graph the following equation: $|x| + |y| = 4$.

209. Determine whether the function $f(x) = \log(\sqrt{9\tan^2 x + 1} + 3\tan x)$ is even, odd or neither.

210. Let $f(x) = 4 - 15x^2 - 2x^3$ and $g(x) = 2x^3 + 18x + 1$. At what value of x is $\dfrac{f'(x)}{g'(x)} > 0$?

211. Write the number 10 as a sum of two positive addends in such a way, that the sum of half of the square of the first addend and the cube of the second addend is a minimum.

212. Solve the inequality $f'(x) + g'(x) < 0$, where $f(x) = \dfrac{x^2 - 1}{x - 3}$ and $g(x) = \dfrac{1}{x - 3}$.

213. Find the domain of the function $f(x) = (x + 0.5)^{\log_{x+0.5} \frac{x^2+2x-3}{4x^2+4x-3}}$.

214. Find the domain of the function $f(x) = \sqrt{4 - x^2} + \arccos\dfrac{1}{x} - \dfrac{1}{x - 2}$.

215. For what value a is the sum of the squares of the roots of the equation $x^2 - ax + a - 2 = 0$ a minimum?

216. Determine whether a polynomial P(x) with integer coefficients exists if
    a) $P(0) = 19$, $P(1) = 85$, $P(2) = 1985$;   b) $P(1) = 19$, $P(19) = 85$.

217. Assume the equation $ax^2 + bx + c = 0$ does not have real roots and $a + b + c < 0$. What is the sign of c ?

218. Among all rectangles inscribed in a semicircle in such a way that a side of the rectangle lies on a diameter, find the rectangle of maximum area.

219. In a given sphere of radius R, inscribe a cylinder of maximum volume.

220. In a given sphere of radius R, inscribe a cone of maximum volume.

221. At what value of a do the polynomials $f(x) = x^4 + ax^2 + 1$ and $g(x) = x^3 + ax + 1$ have a common zero ?

222. At what value of a is the area between the parabola $y = 2x^2$ and the straight line $y = a$ equal to $\dfrac{20\sqrt{30}}{2}$?

223. A tourist is jogging from point A to point B. Point A is located on a straight road and point B is 8 miles from this road. The distance between A and B is 17 miles. At what point does the tourist have to leave the road and turn toward B, so as to minimize the time of jogging, if he can jog 5 miles per hour on the road and 3 miles per hour off the road?

224. There are 25 pupils in a class. Among them 17 pupils can ride a bicycle, 13 can swim and 8 can ski. No pupil can perform all three sports, but the cyclists, swimmers, and skiers have all received grades of A or B in mathematics. This is interesting since 6 pupils in the class have received grades of C or D in mathematics. How many pupils have received a grade of F? How many swimmers can ski?

225. In the picture below there are 7 circles in a triangle. Place in each circle a different number from 1 to 7 in such a way, that the sum of the numbers along each straight line containing 3 circles is the same.

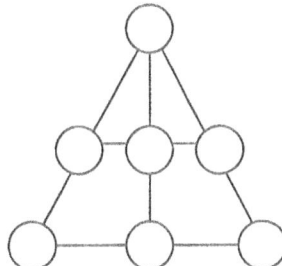

226. The sum of the third and ninth terms of an arithmetic sequence is equal to the minimum value of the trinomial $2x^2 - 4x + 10$. Find the sum of the first eleven terms of this sequence.

227. There are 70 balls in a box: 20 red, 20 yellow, 20 green, and the rest are white and black. The balls are distinct only in color. What is the minimum number of balls that we have to select arbitrarily from the box, in order to obtain at least 10 balls of the same color?

228. Solve the equation $\log \sin(x + |x|) = 0$.

229. Two friends Tom and John live in villages A and B which are 36 miles apart. One day Tom with his dog and John simultaneously leave their homes and start to walk toward each other. Tom walks at the speed of 4 miles per hour. John walks at the speed of 5 miles per hour. The dog runs at the speed of 6 miles per hour. After meeting John, the dog immediately turns back and continues running toward Tom. After meeting Tom, the dog turns back running to John and repeats running between Tom and John until all three of them meet. What distance was covered by the dog?

230. The numbers $5x - y$, $2x + 3y$ and $x + 2y$ form an arithmetic sequence. The numbers $(y + 1)^2$, $xy + 1$ and $(x - 1)^2$ form a geometric sequence. Find x and y.

231. Four dogs A, B, C and D are standing in the corners of a square meadow and suddenly begin to pursue one another as shown by the arrows below. Each dog is running directly after his neighbor - A after B, B after C, C after D, and D after A.

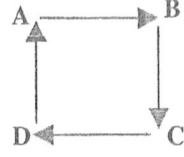

The side of the meadow is 100 meters, and the speed of each dog is 10 meters/sec. After what period of time will the dogs meet one another? Will their tracks ever cross, and where? How long will their tracks be?

232. For which x are the numbers $\log 2$, $\log(2^x - 1)$, $\log(2^x + 3)$ three consecutive terms of an arithmetic sequence?

233. For which α are $\dfrac{\sin\alpha}{6}$, $\cos\alpha$, and $\tan\alpha$ the consecutive terms of a geometric sequence?

234. Solve the equation $3^x + 4^x = 5^x$.

235. Two towns A and B are located 10 miles apart on the shore of a river. Which would take longer for a ship: to sail from A to B and back, or to sail 20 miles in still water?

236. There are two sizes of containers, 130 kilograms and 160 kilograms. Workers have to load 3 tons on a truck using only these containers. Is it possible to do this?

237. It is known that the percent of blonds among people with blue eyes is more than the percent of blonds among all people. Which is larger: the percent of people with blue eyes among blonds, or the percent of people with blue eyes among all people?

238. In a triangle with an angle of 120°, the lengths of the sides form an arithmetic sequence, the difference of which is 1 in. Find the sides of the triangle.

239. Write any integer n using the number 2 three times and mathematical signs and symbols.

240. How many different 3-digit numbers can be written using the digits 0, 1, 2, 3, 4, 5?

241. A hockey team contains 3 attackers, 2 defenders and 1 goalkeeper. If a coach has 6 attackers, 4 defenders and 2 goalkeepers, how many teams is he able to create?

242. One of the legs of a right triangle has a length of 13 in. Find the other sides of this triangle, if they both have lengths expressed by natural numbers.

243. After each meeting of the Mathematical Club some members visit an ice cream café (more than one member but not all of them). There is a very strong rule in this club: after each visit no two participants of this visit can eat ice cream together in the future. At the last meeting, the members of the club recognize that now each member can only eat ice cream alone. Find how many meetings the club members could have if the club contains 4 members. (Give all possible answers)

244. One quarter of all participants in a chess tournament were grandmasters and the rest were masters. Each two participants played one game with each other. For winning a participant received 1 point, for a draw 0.5 point and for losing 0 points. The sum of the points earned by masters was 1.2 times the sum of the points earned by grandmasters. How many masters and grandmasters participated in this tournament?

245. Solve this cryptarithm

$$\frac{EVE}{DID} = .TALKTALKTALK...$$

The same letters stand for the same digits, zero included. The fraction EVE/DID has been reduced to its lowest term. Its decimal form has a repeating period of four digits.

246. The area of an equilateral triangle constructed on the hypotenuse of a right triangle is twice the area of the right triangle. Find the ratio of the legs of the right triangle.

247. Solve the following system of equations:

$$\begin{cases} x^2 + y^2 = 2 \\ x^3 + y^3 = 2 \end{cases}$$

248. Three candles have the same length but different thicknesses. The first candle was lighted 1 hour earlier than the other two, which were lighted simultaneously. At a certain moment the first and third candles had the same length, and 2 hours later the first and second candles had the same length. How long will it take the first candle to burn completely, if the second burns 12 hours and the third burns 8 hours?

249. Evaluate without using a calculator : $\tan 5° \cdot \tan 55° \cdot \tan 65° \cdot \tan 75°$.

250. A student noticed that the lengths of three sides of a triangle and its area are four consecutive natural numbers. What are the lengths of the sides of this triangle?

ANSWERS, HINTS, SOLUTIONS

NUMBERS , SEQUENCES , POLYNOMIALS

1. $p^2 - 1 = (p + 1)(p - 1)$ is a product of two consecutive even numbers. One of them is divisible by 2 and the other by 4. $(p-1)$, p ,$(p + 1)$ are 3 consecutive numbers. One of them is divisible by 3. p is a prime number, so either $(p + 1)$ or $(p - 1)$ is divisible by 3. Thus, $(p + 1)(p - 1)$ is divisible by $2 \cdot 3 \cdot 4 = 24$.

2. 24

3. 648. (**Hint:** Solve the equation $2a^2 = 3b^3$ for natural numbers a and b).

4. 2178

5. 89 (**Hint:** Write the number in the form $10x + y$.)

6. 36

7. a)
$$\begin{array}{r} 9567 \\ + 1085 \\ \hline 10652 \end{array}$$

   b)
$$\begin{array}{r} 9567 \\ + 1085 \\ \hline 10652 \end{array}$$

8. a)
$$\begin{array}{r} 134 \\ \times 87 \\ \hline 938 \\ 1072 \\ \hline 11658 \end{array}$$

   b)
$$\begin{array}{r} 31245 \\ \times 237 \\ \hline 218715 \\ 93735 \\ 62490 \\ \hline 7405065 \end{array}$$

   c)
$$\begin{array}{r} 90809 \\ \hline 12\,|\,1089708 \\ 108 \\ \hline 97 \\ 96 \\ \hline 108 \\ 108 \\ \hline \end{array}$$

9. **Hint:** Factor the expression as a difference of squares and evaluate the last digits.

10. $2^{2002} - 1 = (2^{1001} + 1)(2^{1001} - 1)$. Both factors are consecutive odd numbers. 2^{1001} is an even number not divisible by 3. So either the preceding or the following number is divisible by 3.

11. $n^{12} - n^8 - n^4 + 1 = n^8(n^4 - 1) - (n^4 - 1) = (n^4 - 1)(n^8 - 1) =$
    $= (n^2 + 1)(n^4 + 1)(n + 1)(n - 1)(n^2 + 1)(n + 1)(n - 1)$. All factors are even. Either $n - 1$ or $n + 1$ are divisible by 4. So the number is divisible by $2^9 = 512$.

12. **Hint:** The expression is the product of five consecutive even numbers. So at least one of them is divisible by 5, one by 3, three of them by 2, and two of them by 4. $5 \cdot 3 \cdot 8 \cdot 16 = 1920$.

13. If either m or n is divisible by 3, then mn is divisible by 3. If neither m nor n is divisible by 3, then we have 3 possible cases: 1) $m = 3k + 1$; $n = 3p + 1$, and $m - n$ is divisible by 3; 2) $m = 3k + 2$; $n = 3p + 2$, and $m - n$ is divisible by 3; 3) $m = 3k + 1$; $n = 3p + 2$, and $m + n$ is divisible by 3. (The case $m = 3k + 2$; $n = 3p + 1$ is analogous to case 3).

14. 43

15. The expression is the product of five consecutive natural numbers. At least one of them is divisible by 5, two by 3, one by 2 and two by 4. So the number is divisible by $5 \cdot 9 \cdot 2 \cdot 16 = 1440$.

16. Each addend is odd, so the number is even. After dividing 43^{17} by 3 the remainder is 1 and the remainder of 17^{17} is 2, so the sum is divisible by 3. After dividing 43^{17} by 5 the remainder is 3 and the remainder of 17^{17} is 2, so the sum is divisible by 5. Thus the number is divisible by $2 \cdot 3 \cdot 5 = 60$. An alternative solution is based on the fact that $a^n + b^n$ is divisible by $(a + b)$ for any odd n.

17. $\dfrac{n}{12} + \dfrac{n^2}{8} + \dfrac{n^3}{24} = \dfrac{n(n+1)(n+2)}{24}$. The numerator contains three consecutive integers. One of them is divisible by 2, one of them by 4, and one of them by 3. Thus the number is divisible by 24.

18. $3 \cdot 4 \cdot 5 \cdot 7 + 1 = 421$

19. Let the number be $100x + 10y + z$. Then the difference is $(100x + 10y + z) - (100z + 10y + x) = 99(x - z) = 9 \cdot 11(x - z)$. $x - z$ cannot be equal to 11. Thus the number cannot be a perfect square.

20. For 2 jumps the dog covers 4 meters and for 3 jumps the fox covers 3 meters. So for one cycle the distance between them is decreasing by 1 meter. Thus the dog has to run $4 \cdot 50 = 200$ meters to catch up to the fox.

21. $1^2 - 2^2 + 3^2 - 4^2 + 5^2 - 6^2 + \ldots + 999^2 - 1000^2 =$
$= (1 + 2)(1 - 2) + (3 + 4)(3 - 4) + (5 + 6)(5 - 6) + \ldots + (999 + 1000)(999 - 1000) =$
$= (-1)(3 + 7 + 11 + \ldots + 1999) = \dfrac{(3 + 1999) \cdot 500}{2}(-1) = -500500.$

22. The numbers $a, a + d, a + 2d$ satisfy the condition $(3a + 3d) \cdot 5 = a(a + d)(a + 2d)$. Then $a(a + 2d) = 15$. So $a = \pm 1$ and $d = \pm 7$; $a = \pm 3$ and $d = \pm 1$; $a = \pm 5$ and $d = \pm 1$; $a = \pm 15$ and $d = \pm 7$.
Thus the answer: 3,4,5; 5,4,3; -3,-4,-5; -5,-4,-3; 1,8,15; 15,8,7; -1,-8,-15; -15,-8,-1.

23. $\sqrt{5\sqrt{3\sqrt{5\sqrt{3\sqrt{5\sqrt{3\ldots}}}}}} = 5^{\frac{1}{2}} \cdot 3^{\frac{1}{4}} \cdot 5^{\frac{1}{8}} \cdot 3^{\frac{1}{16}} \cdot \ldots = 5^{\frac{1}{2}+\frac{1}{8}+\frac{1}{32}+\ldots} \cdot 3^{\frac{1}{4}+\frac{1}{16}+\frac{1}{64}+\ldots} = 5^{\frac{2}{3}} \cdot 3^{\frac{1}{3}} = \sqrt[3]{75}.$

$\dfrac{1}{2} + \dfrac{1}{8} + \dfrac{1}{32} + \ldots = \dfrac{\frac{1}{2}}{1 - \frac{1}{4}} = \dfrac{2}{3};$ $\qquad$ $\dfrac{1}{4} + \dfrac{1}{16} + \dfrac{1}{64} + \ldots = \dfrac{\frac{1}{4}}{1 - \frac{1}{4}} = \dfrac{1}{3}.$

24. $a_6 + a_9 + a_{12} + a_{15} = a_1 + 5d + a_1 + 8d + a_1 + 11d + a_1 + 14d = 4a_1 + 38d = 20;$

$2a_1 + 19d = 10; \quad a_1 + a_2 + a_3 + \ldots + a_{20} = \dfrac{(2a_1 + 19d)20}{2} = 10 \cdot 10 = 100.$

25. The solution of the system of equations

$\left\{ \begin{array}{l} \dfrac{b_1}{1-r} = 15 \\[4mm] \dfrac{b_1^{\,2}}{1-r^2} = 75 \end{array} \right.$ is $b_1 = 7.5; \ r = \dfrac{1}{2}.$

The sum of the first 6 terms is

$S_6 = \dfrac{15\left(\left(\dfrac{1}{2}\right)^6 - 1\right)}{2\left(\dfrac{1}{2} - 1\right)} = \dfrac{945}{64}.$

26. $1 + 11 + 111 + \ldots + \underbrace{111\ldots1}_{} = (1 + 10 + 100 + \ldots + \underbrace{1000\ldots0}_{n}) + (1 + 10 + 100 + \underbrace{1000\ldots0}_{n-1}) +$

$+ (1 + 10 + 100 + \ldots \underbrace{1000\ldots0}_{n-2}) + \ldots + (1 + 10) + 1 =$

$\dfrac{1(10^n - 1)}{10 - 1} + \dfrac{1(10^{n-1} - 1)}{10 - 1} + \dfrac{1(10^{n-2} - 1)}{10 - 1} + \ldots + \dfrac{10^2 - 1}{10 - 1} + 1 = \dfrac{1}{9}\left(\dfrac{10^{n+1} - 10}{9} - n\right) = \dfrac{1}{81}\left(10^{n+1} - 10 - 9n\right)$

27. $S = 1 + 3x + 5x^2 + 7x^3 + 9x^4 + \ldots = (1 + x + x^2 + x^3 + x^4 + \ldots) + 2x(1 + 2x + 3x^2 + 4x^3 + \ldots) =$

$= \dfrac{1}{1-x} + 2x\left((1 + x + x^2 + x^3 + \ldots) + (x + x^2 + x^3 + x^4 + \ldots) + (x^2 + x^3 + x^4 + x^5 + \ldots) + \ldots\right) =$

$= \dfrac{1}{1-x} + \dfrac{2x}{1-x}(1 + x + x^2 + x^3 + x^4 + \ldots) = \dfrac{1}{1-x} + \dfrac{2x}{1-x} \cdot \dfrac{1}{1-x} = \dfrac{1+x}{(1-x)^2}. \quad S\left(\dfrac{3}{5}\right) = 10.$

28. $\dfrac{244 \cdot 395 - 151}{244 + 395 \cdot 243} = \dfrac{(243+1)395 - 151}{244 + 395 \cdot 243} = \dfrac{243 \cdot 395 + 395 - 151}{395 \cdot 243 + 244} = \dfrac{243 \cdot 395 + 244}{395 \cdot 243 + 244} = 1.$

29. Let the number be $10x + y$. Then the desired sum is $10x + y + 10y + x = 11x + 11y = 11(x + y)$. To be a perfect square, $x + y$ must be 11 and $x \le 9; \ y \le 9$. Thus the numbers are 29, 38, 47, 56, 65, 74, 83, 92.

30. $6\dfrac{4}{11}$ min.

31. An odd number can be written in the form $2n - 1$ where n is a natural number. $(2n - 1)^2 = 4n^2 - 4n + 1 = 4n(n - 1) + 1$. Either n or $n - 1$ is even. So $4n(n - 1)$ is divisible by 8. Obviously the remainder is 1.

32. $2a^2 + 2b^2 = (a + b)^2 + (a - b)^2.$

33. $(n - 1)n(n + 1) + n = n^3 - n + n = n^3.$

34. 0

35. **Hint:** Multiply the third factor by the fourth. Then multiply the result by the second factor, and then by the first factor.

36. $-2.$ (**Hint:** $9 + 4\sqrt{5} = (2 + \sqrt{5})^2.$)

37. If $a + b + c = 0$, then $a + b = -c$. Take the cube of the left and right parts of the equality: $(a + b)^3 = (-c)^3 \Rightarrow a^3 + 3ab(a + b) + b^3 = -c^3 \Rightarrow a^3 - 3abc + b^3 = -c^3 \Rightarrow a^3 + b^3 + c^3 = 3abc.$

38. $1 + 3^x + 9^x = 1 + 3^{3k+1} + 9^{3k+1} = 1 + 3 \cdot 27^k + 9 \cdot 27^{2k}$. The remainder after dividing 27 by 13 is 1, which also is the remainder if you divide (27 to any power) by 13. So the remainder after dividing this expression by 13 is $1 + 3 + 9 = 13$, i.e. the expression is divisible by 13.

39. $(n - 1)^3 + n^3 + (n + 1)^3 = 3[(n - 1)n(n + 1) + 3n]$. The expression within brackets is divisible by 3, so the whole expression is divisible by 9.

40. $n^3 + 11n = n^3 - n + 12n = (n - 1)n(n + 1) + 12n$. At least one of three consecutive numbers is even and at least one of them is divisible by 3. So the expression is divisible by 6.

41. 7. (**Hint:** Compute the square of the unknown expression.)

42. If x is the number of hundreds, y is the number of tens and z is the number of units, then the number is $100x + 10y + z$. The difference of this number and the number written in the opposite direction is $99(x - y)$. Check that if $1 \leq x - y \leq 9$, then the sum of the digits of the difference is 18.

43. Separate all coins into 3 parts : 27, 27, and 26. Place on the scale 27 and 27. If one of the sides is lighter, then the false coin is in this group. If there is an equilibrium, then the false coin is among the group of 26. In the first case the following scheme allows us to find the false coin:
a) 9, 9, 9; b) 3, 3, 3); c) 1, 1, 1. In the second case we separate coins into groups of 9, 9, 8 and proceed by analogy with the explanation at the beginning. In both cases we can find the false coin in 4 weighings.

44. Yes. Check that $9^2 + 12^2 = 15^2$. So if you take 9 matches and 12 matches for the legs and 15 for the hypotenuse, the triangle will be right.

45. $5^{5k+1} + 4^{5k+2} + 3^{5k} = 5 \cdot (5^5)^k + 4^2 \cdot (4^5)^k + (3^5)^k$. Check that $5^5 \equiv 1 (\text{mod} 11)$, $4^5 \equiv 1 (\text{mod} 11)$, and $3^5 \equiv 1 (\text{mod} 11)$. Then the remainder after dividing the first addend by 11 is 5, the second addend by 11 is 5 and the third by 11 is 1. Thus the number is divisible by 11.

46. Note that $a^n + b^n$ is divisible by $a + b$ if n is an odd number (look at formula (7)). Thus the number

$$3^{105} + 4^{105} = \left(3^3\right)^{35} + \left(4^3\right)^{35} = \left(3^5\right)^{21} + \left(4^5\right)^{21} = \left(3^7\right)^{15} + \left(4^7\right)^{15} \quad \text{is divisible by } 3^3 + 4^3 = 7 \cdot 13,$$

by $3^5 + 4^5 = 7 \cdot 181$, and by $3^7 + 4^7 = 49 \cdot 379$.

Check that $3^{105} \equiv 3(\bmod 5)$ and $4^{105} \equiv -1(\bmod 5)$, and consequently $3^{105} + 4^{105} \equiv 2(\bmod 5)$,

which means that the number $3^{105} + 4^{105}$ is not divisible by 5.

Similarly $3^{105} \equiv 1(\bmod 11)$, $4^{105} \equiv 1(\bmod 11)$, and $3^{105} + 4^{105} \equiv 2(\bmod 11)$,

which shows that $3^{105} + 4^{105}$ divided by 11 gives the remainder 2.

47. 1994 pages.

48.

$$\frac{1}{1 \cdot 2} + \frac{1}{2 \cdot 3} + \frac{1}{3 \cdot 4} + \ldots + \frac{1}{99 \cdot 100} = \left(\frac{1}{1} - \frac{1}{2}\right) + \left(\frac{1}{2} - \frac{1}{3}\right) + \left(\frac{1}{3} - \frac{1}{4}\right) + \ldots + \left(\frac{1}{99} - \frac{1}{100}\right) = \frac{99}{100}.$$

49. $\dfrac{1}{\log_2 \pi} + \dfrac{1}{\log_5 \pi} = \log_\pi 2 + \log_\pi 5 = \log_\pi 10 > 2.$

50. Yes. $203 = 7 + 29 + \underbrace{1 + 1 + 1 + \ldots + 1}_{167} = 7 \cdot 29 \cdot \underbrace{1 \cdot 1 \cdot 1 \cdot \ldots \cdot 1}_{167}.$

51. 13. (**Hint:** Calculate some terms at the beginning of the sequence and notice the periodicity.)

52. a) $3^{500} = (3^5)^{100} = 243^{100};\ 7^{300} = (7^3)^{100} = 343^{100}.$ Thus $3^{500} < 7^{300}$.

b) Let $\log_5 6 = m$ or $5^m = 6$. Then $\log_5 6 = \log_{5^m} 6^m = \log_6 6^m$.

But $6^m = (5 + 1)^m > 5^m + 1^m \Rightarrow 6^m > 7$. Thus, $\log_5 6 > \log_6 7$.

(We used the fact that for $n \geq 1$, $(a + b)^n \geq a^n + b^n$.)

53. Obviously the price of the pen is \$1 and some cents. Let x be the number of cents. Then $200 \leq 9x < 300$ and $23 \leq x < 33$.
On the another hand, $200 \leq 13x < 300$ and $16 \leq x \leq 23$. Thus $x = 23c$.
So the price of the pen is \$1.23.

54. **Hint:** Apply to the formulas in (5) for factoring $n^3 + 1$ and $n^3 - 1$.

55. 2, 5, 7.

56. No. Let us assume that the legs of a right triangle a and b both are odd numbers. Then $a = 2k + 1$, $b = 2m + 1$ and the hypotenuse c must be an even number: $c = 2n$. Check that The Pythagorean Theorem $(2k + 1)^2 + (2m + 1)^2 = (2n)^2$ does not work in such case, because the right side of this equality is divisible by 4 whereas the left side is not.

57. Note that for any natural n the sum of the digits in the numerator is 9.

58. $a^{\log_c b} - b^{\log_c a} = a^{\log_c b} - \left(a^{\log_a b}\right)^{\frac{\log_b a}{\log_b c}} = a^{\log_c b} - a^{\frac{1}{\log_b c}} = a^{\log_c b} - a^{\log_c b} = 0$.

59. (**Hint:** Transform each logarithm to the base 2.)

60. 48, 49, 50. (Try to find three other numbers.)

61. **a)** $x^8 + x^4 + 1 = x^8 + 2x^4 + 1 - x^4 = (x^4 + 1)^2 - x^4 = (x^4 + 1 - x^2)(x^4 + 1 + x^2) =$
$(x^4 - x^2 + 1)(x^4 + 2x^2 + 1 - x^2) = (x^4 - x^2 + 1)((x^2 + 1)^2 - x^2) =$
$(x^4 - x^2 + 1)(x^2 + x + 1)(x^2 - x + 1)$.

b) $x^5 + x + 1 = (x^5 + x^4 + x^3) - (x^4 + x^3 + x^2) + (x^2 + x + 1) =$
$x^3(x^2 + x + 1) - x^2(x^2 + x + 1) + (x^2 + x + 1) = (x^3 - x^2 + 1)(x^2 + x + 1)$.

62. $x^9 + x^4 - x - 1 = (x^9 - 1) + x(x^3 - 1) = (x^3 - 1)(x^6 + x^3 + 1 + x) =$
$(x - 1)(x^2 + x + 1)[x^3(x^3 + 1) + (x + 1)] = (x - 1)(x + 1)(x^2 + x + 1)(x^5 - x^4 + x^3 + 1) =$
$(x - 1)(x + 1)(x^2 + x + 1)[x^3(x^2 + 1) + (1 - x^2)(1 + x^2)] =$
$(x - 1)(x + 1)(x^2 + 1)(x^2 + x + 1)(x^3 - x^2 + 1)$.

63. If $p = 3$, then $8p^2 + 1 = 73$ which is a prime number. Now we have to prove that for a prime $p \neq 3$, the number $8p^2 + 1$ is composite. Among all numbers divisible by 3, only 3 is a prime. Let p not be divisible by 3.
Then $p = 3k \pm 1$, $8p^2 + 1 = 8(3k \pm 1)^2 + 1 = 72k^2 \pm 48k + 9 = 3(24k^2 \pm 16k + 3)$.
This number obviously is divisible by 3, i.e. is composite. So $8p^2 + 1$ is a prime number only for $p = 3$.

64. Let x and y be unknown numbers. Then the given information can be described by the following system of equations:
$$\begin{cases} x + y = 15 \\ \dfrac{x + y}{2} = \dfrac{5}{4}\sqrt{xy} \end{cases}. \qquad \text{Answer: } x = 12, \ y = 3.$$

65. Answer: $\dfrac{2 + \sqrt{2} - \sqrt{6}}{4}$. (**Hint:** Multiply the numerator and denominator of the given fraction by $(\sqrt{3} + \sqrt{2} - 1)$ and then by $(2 - \sqrt{6})$.)

66. 0.

67. $a < b$. (**Hint:** Rationalize the denominators and then try to eliminate radicals by comparing the squares of the numbers.)

68.
$$\sqrt{\underbrace{111...1}_{2n} - \underbrace{222...2}_{n}} = \sqrt{\underbrace{111...1}_{n} \cdot (\underbrace{100...01}_{n+1} - 2)} = \sqrt{\underbrace{111...1}_{n} \cdot \underbrace{999...9}_{n}} = \sqrt{9 \cdot \underbrace{111...1}_{n}^2} = \underbrace{333...3}_{n}.$$

69. Suppose we have the contrary statement, e.g. $x^2 + 3x + 5 = 121k$ at some x where k is an integer. Then one of the solutions of the equation $x^2 + 3x - (121k - 5) = 0$ is an integer. This solution
$$x = \frac{-3 \pm \sqrt{9 + 484k - 20}}{2} = \frac{-3 \pm \sqrt{11(44k - 1)}}{2}$$ can be an integer if and only if
$11(44k - 1)$ is a perfect square, but this cannot be true for any k.

70. Notice that $x^2(x^2 - x + 1)^2 \neq 0$. After dividing each term of the original equation by this expression we will have the equivalent equation:

$$\frac{(x^2 - x + 1)^2}{x^2} - 10 + \frac{9x^2}{(x^2 - x + 1)^2} = 0.$$ Using the new variable $y = \left(\frac{x^2 - x + 1}{x^2}\right)^2$ and

solving the equation $y - 10 + \frac{9}{y} = 0$ we have $y = 1$ or $y = 9$.

Returning to the variable x, we have 4 real solutions: $x = \pm 1$, $x = 2 \pm \sqrt{3}$.

71. Notice that $2 - \sqrt{3} = \frac{1}{2 + \sqrt{3}}$. Let a new variable y be $\left(\sqrt{2 + \sqrt{3}}\right)^x$. Then the equation

for y is $y + \frac{1}{y} = 4$. The solution is $y = 2 \pm \sqrt{3}$. Solving the equation for x

we have $x = \pm 2$.

72. Introduce the new variable $y = \sqrt{a + x}$. Then the original equation at $x \geq 0$ and $y \geq 0$ is equivalent to the system:

$$\begin{cases} \sqrt{a - y} = x \\ \sqrt{a + x} = y \end{cases} \Rightarrow \begin{cases} a - y = x^2 \\ a + x = y^2 \end{cases}$$

After subtracting the first equation from the second we have $(y + x)(y - x) = y + x$. So either $x = -y$ (possible only if $x = y = a = 0$) or $y - x = 1$. The latter case gives the

equation $\sqrt{a + x} = x + 1$. The solution is $x = \frac{-1 + \sqrt{4a - 3}}{2}$ at $a \geq 1$.

73. Write the equation in the form $x^2 - 3x + 11 + 2\sqrt{x^2 - 3x + 11} = 15$. Substituting

$y = \sqrt{x^2 - x + 11}$, $y \geq 0$ and solving the equation $y^2 + 2y - 15 = 0$ we obtain $y = 3$. The solution of the original equation is $x = 1$ or $x = 2$.

74. $x^{\frac{2}{3}} \cdot x^{\frac{2}{9}} \cdot x^{\frac{2}{27}} \cdot \ldots = 49 \Rightarrow x^{\frac{2}{3} + \frac{2}{9} + \frac{2}{27} + \ldots} = x^{\frac{\frac{2}{3}}{1 - \frac{1}{3}}} = x^1 = 49$.

75. $ax^2 + 2ax + 0.5 \geq 0$ for $a > 0$ only if the discriminant $D \leq 0$.

For $ax^2 + 2ax + 0.5 = 0$ $x = \frac{-a \pm \sqrt{a^2 - 0.5a}}{a}$. So $a(a - 0.5) \leq 0$ and $a > 0$.

The answer is $0 < a \leq 0.5$.

76. $x^2 - 4x \geq 0$ at $x \leq 0$ or $x \geq 4$. For all $x \leq 0$ the original inequality is valid. For $x \geq 4$ we have to solve the inequality $x^2 - 4x > (x - 3)^2$, the solution of which is $x > 4.5$. Thus $x \leq 0$ or $x > 4.5$.

77. Transforming exponential expressions to the same base we have $3^{2[\log(x-a)-\log 2]} = 3^{\log(x-1)} \Rightarrow$

$2[\log(x-a)-\log 2] = \log(x-1) \Rightarrow \log\dfrac{(x-a)^2}{4} = \log(x-1) \Rightarrow (x-a)^2 = 4(x-1)$.

The solution of this quadratic equation is $x = a + 2 + 2\sqrt{a}$, where $0 \le a \le 1$.

78. $\dfrac{9}{16}$

79. The equivalent equation is $9^{x-1} + 7 = (3^{x-1} + 1)^2$. Then $9^{x-1} + 7 = 9^{x-1} + 2 \cdot 3^{x-1} + 1$; $2 \cdot 3^{x-1} = 6$; $3^{x-1} = 3$; $x - 1 = 1$; $x = 2$.

80. **Hint:** This is a quadratic equation with respect to $\log x$. The answer is $x = 10$.

81. For $x > 0$ the equivalent equation is $\log_4^2 x - \dfrac{1}{2}\log x - 1.5 = 0$. The solution of this quadratic with respect to the $\log x$ equation is $\log_4 x = \dfrac{3}{2}$ or $\log_4 x = -1$. Then $x = 8$ or $x = \dfrac{1}{4}$.

82. $x = 1000$ or $x = 0.01$. (**Hint:** Use a new variable $y = \log x$).

83. $x = 9$. (**Hint:** $\log_{\frac{1}{3}} x = -\log_3 x$)

84. $x = 0$ or $x = 3$.

85. Using the properties of logarithms we have $\log_2 \dfrac{25^{x+3} - 1}{5^{x+3} + 1} = 2$ and then $\dfrac{25^{x+3} - 1}{5^{x+3} + 1} = 4$;

$25^{x+3} - 4 \cdot 5^{x+3} - 5 = 0$. If $y = 5^{x+3}$, then $y^2 - 4y - 5 = 0$ and $y = 5$ or $y = -1$. $5^{x+3} = -1$ does not have real solutions. The solution of $5^{x+3} = 5$ is $x = -2$.

86. $x = 2$. (Use the properties of logarithms to obtain an exponential equation.)

87. Transforming to the same bases on both sides of the equation we have

$7^{-x^2 - x + 0.5} = 7^{-0.5}$; $2x^2 + x - 1 = 0$; $x = \dfrac{1}{2}$ or $x = -1$.

88. Notice that $3^{x+1} = 3 \cdot 3^x$. Then the original equation can be written in the form $6 \cdot 3^x - 3^x = 15$; $5 \cdot 3^x = 15$; $3^x = 3$; $x = 1$.

89. $x = -1$.

90. $x = 1$ or $x = 2$. (**Hint:** $3^{3-x} = \dfrac{27}{3^x}$; then use $y = 3^x$).

91. $x = 3$ or $x = 11$. (**Hint:** Use a new variable $y = 2^{\sqrt{x-2}}$, then $4^{\sqrt{x-2}} = y^2$.)

92. $x = 2$.

93. The solution of the equivalent system $\begin{cases} x + y = \dfrac{1}{9} \\ x - y = 9 \end{cases}$

   is $x = \dfrac{41}{9}$, $y = -\dfrac{40}{9}$.

94. Using properties of logarithms we have the new system

   $\begin{cases} \dfrac{x^2 + y^2}{13} = 10 \\ \dfrac{x+y}{x-y} = 8 \end{cases}$ By substituting $x = \dfrac{9y}{7}$ from the second equation in the first

   we have $y^2 + \dfrac{81y^2}{49} = 130$ and $y = 7$, $x = 9$ or $y = -7$, $x = -9$.

   To be in the domain of the original equation we choose the answer $x = 9$, $y = 7$.

95. Rewrite the system in the form

   $\begin{cases} 3^y \cdot 3^{2x} = 81 \\ \log \dfrac{(x+y)^2}{x} = \log 9 \end{cases} \Rightarrow \begin{cases} 2x + y = 4 \\ (x+y)^2 = 9x \end{cases} \Rightarrow \begin{cases} y = 4 - 2x \\ (4 - x)^2 = 9x. \end{cases}$

   The solution to the second equation is $x = 1$ or $x = 16$.
   Answer: $(1,2)$, $(16,-28)$.

96. Rewrite the system in the form

   $\begin{cases} 10 \cdot 10^{\log(x+y)} = 50 \\ \log[5(x+y)(x-y)] = 2 \end{cases} \Rightarrow \begin{cases} x + y = 5 \\ 5(x+y)(x-y) = 100 \end{cases} \Rightarrow \begin{cases} x + y = 5 \\ x - y = 4 \end{cases}$

   Answer: $x = 4.5$, $y = 0.5$.

97. The equivalent system is

   $\begin{cases} 3^x \cdot 2^y = 576 \\ y - x = 4 \end{cases} \Rightarrow \begin{cases} 3^x \cdot 2^{4+x} = 576 \\ y = 4 + x \end{cases} \Rightarrow \begin{cases} 16 \cdot 6^x = 576 \\ y = 4 + x \end{cases} \Rightarrow \begin{cases} x = 2 \\ y = 6 \end{cases}$

98. Rewrite the inequality in the form $4^{5-2x} \le 4^{-1}$. By properties of exponential functions
   with a base larger than 1, the equivalent inequality is $5 - 2x \le -1$, i.e. $x \ge 3$.
   Answer: $[3, \infty)$.

99. Rewrite the inequality in the form $0.4^{2x+1} > 0.4^2$. By properties of exponential functions
   with a base less than 1, the equivalent inequality is $2x + 1 < 2$, i.e. $x < \dfrac{3}{2}$.

   Answer: $\left(-\infty, \dfrac{3}{2}\right)$.

100. $\left(-\infty, -\dfrac{3}{5}\right]$

101. Let $y = 2^x$. Then the solution of the inequality $y^2 - 2y - 8 > 0$ is $y < -2$ or $y > 4$. So either $2^x < -2$ or $2^x > 4$. The first inequality does not have a solution, but the solution of the second is $x > 2$. Answer: $(2, \infty)$

102. $(2,3,4)$ or $(-2,-3,-4)$

103. Let $a = xy$, $b = yz$, and $c = xz$. Then the system
$$\begin{cases} a + b = 8 \\ b + c = 20 \\ a + c = 13 \end{cases}$$
has the solution $a = 1$, $b = 7$, $c = 13$.
The new system
$$\begin{cases} xy = 1 \\ yz = 7 \\ xz = 13 \end{cases}$$

has the solutions
$$x = \sqrt{\dfrac{13}{7}}, y = \sqrt{\dfrac{7}{13}}, z = \sqrt{91} \quad \text{or}$$
$$x = -\sqrt{\dfrac{13}{7}}, y = -\sqrt{\dfrac{7}{13}}, z = -\sqrt{91}.$$

104. By addition,
$$\begin{cases} x^2 y = 50 \\ xy^2 = 20. \end{cases}$$
After dividing the first equation by the second and using the method of substitution, we obtain $x = 5$, $y = 2$.

105. Transform the system to the form
$$\begin{cases} (x^2 + y^2)^2 - 2x^2 y^2 = 17 \\ (x^2 + y^2) + 2xy = 9 \end{cases}. \quad \text{Then use new variables } u = x^2 + y^2 \text{ and } v = xy.$$
Answer: $(1,2), (2,1)$.

106. After adding the left and right sides of the equations, an equivalent system is
$$\begin{cases} (x + y)^2 + (x + y) = 2 \\ x + xy + y = -5. \end{cases} \quad \text{Let } u = x + y. \text{ Then solving the first equation}$$
with respect to the variable u, $u^2 + u - 2 = 0$ we have to solve the following systems:
$$\begin{cases} x + y = -2 \\ x + xy + y = -5 \end{cases} \quad \text{or} \quad \begin{cases} x + y = 1 \\ x + xy + y = -5. \end{cases}$$
The solutions of both systems can be found by the method of substitution.
Answer: $(-3,1), (1,-3), (3,-2), (-2,3)$.

107. From the second equation we have that either $x = 0$ or $x^2 - 4y^2 = 0$.
If $x = 0$, then $y = 0$ from the second equation and this solution does not satisfy the first equation. If $x^2 - 4y^2 = 0$, then
$$\begin{cases} x - y = 2 \\ x - 2y = 0 \end{cases} \quad \text{or} \quad \begin{cases} x - y = 2 \\ x + 2y = 0. \end{cases} \quad \text{Answer: } (4,2), \left(1\dfrac{1}{3}, -\dfrac{2}{3}\right)$$

108. If $a = 0$, then $x = y = c$ where $c \in R$. If $a \neq 0$, then

$$\begin{cases} x^2 = (x-a)y \\ y^2 - xy = 9ax \end{cases} \Leftrightarrow \begin{cases} x^2 - xy = -ay \\ y^2 - xy = 9ax \end{cases} \Leftrightarrow \begin{cases} x(y-x) = ay \\ y(y-x) = 9ax \end{cases}.$$

After dividing the second equation by the first, we obtain $\dfrac{y}{x} = \dfrac{9x}{y}$, i.e. $y = \pm 3x$.

Then by substituting we have $x = \dfrac{3}{2}a$, $y = \dfrac{9}{2}a$ or $x = \dfrac{3}{4}a$, $y = -\dfrac{9}{4}a$.

109. One obvious solution of the given system is $x = 0$, $y = 0$. Now let $x \neq 0$, $y \neq 0$. Then after dividing the second equation by the first, we have the equivalent system

$$\begin{cases} 2(x+y) = 5xy \\ 4(x^2 - xy + y^2)\,13 \end{cases} \Leftrightarrow \begin{cases} 2(x+y) = 5xy \\ 4((x+y)^2 - 3xy) = 13 \end{cases}.$$

Let $x + y = u$ and $xy = v$. The new system with respect to the new variables is

$$\begin{cases} 2u = 5v \\ 4(u^2 - 3v) = 13 \end{cases}.$$

This system has two solutions: $v = 1$, $u = \dfrac{5}{2}$ or $v = -\dfrac{13}{25}$, $u = -\dfrac{13}{10}$.

To return to the initial variables we have to solve two systems:

$$\begin{cases} x + y = \dfrac{5}{2} \\ x\,y = 1 \end{cases} \quad \text{or} \quad \begin{cases} x + y = -\dfrac{13}{10} \\ xy = -\dfrac{13}{25} \end{cases}.$$

Answer: $(0,0)$, $\left(\dfrac{1}{2},2\right)$, $\left(2,\dfrac{1}{2}\right)$, $\left(\dfrac{-13+\sqrt{377}}{20}, \dfrac{-13-\sqrt{377}}{20}\right)$, $\left(\dfrac{-13-\sqrt{377}}{20}, \dfrac{-13+\sqrt{377}}{20}\right)$.

110. Introduce new variables $u = x^2 - y^2$, $v = xy$. Then

$$\begin{cases} u^2 + v^2 = 13 \\ u + 2v = 1 \end{cases} \Leftrightarrow \begin{cases} 1 - 4v + 4v^2 + v^2 = 13 \\ u = 1 - 2v \end{cases}.$$

The solution of this system is $u = -3$, $v = 2$ or $u = \dfrac{17}{5}$, $v = -\dfrac{6}{5}$.

Answer:

$(-1,-2)$, $(1,2)$, $\left(\dfrac{\sqrt{10(\sqrt{433}+17)}}{10}, -\dfrac{\sqrt{10(\sqrt{433}-17)}}{10}\right)$, $\left(-\dfrac{\sqrt{10(\sqrt{433}+17)}}{10}, \dfrac{\sqrt{10(\sqrt{433}-17)}}{10}\right)$.

111. The system of linear equations has infinitely many solutions if and only if $\dfrac{k-2}{2} = \dfrac{27}{k+1} = \dfrac{4.5}{-1}$. Only $k = -7$ satisfies this condition.

112. Let x be the age of the father and y be the age of the son now. Then the situation can be described by the following system of equations:

$$\begin{cases} x - 22 = 0.5(22 - y) \\ x = 5y \end{cases}$$ the solution of which is $x = 30$, $y = 6$.

113. Denote by x meters per minute the speed of the first bicyclist and by y meters per minute the speed of the second. Then the two conditions of the problem can be described by the system of equations

$$\left\{\begin{array}{l} 24y - 24x = 800 \\ 3x + 3y = 800 \end{array}\right.$$

The solution is $\quad x = 150\dfrac{m}{min} = 9\dfrac{km}{h}, \quad y = \dfrac{350}{3}\dfrac{m}{min} = 7\dfrac{km}{h}$.

114. Let x be the number of days to finish the job according to plan. Then the following equation reflects the situation:

$$\left(\frac{360}{x} + 4\right)(x - 1) = 360.$$

The solution is $x = 10$. Answer: 9 days.

115. Let the speed of the first student be x steps per minute and the speed of the other student be 2x steps per minute. Then the first walked down $\dfrac{40}{x}$ minutes and the second $\dfrac{60}{2x} = \dfrac{30}{x}$ minutes. Let y steps per minute be the speed of the escalator.

Then the length of the escalator L can be computed twice:

$L = 40 + \dfrac{40}{x}\cdot y \quad$ or $\quad L = 60 + \dfrac{30}{x}\cdot y$. The equation $\quad 40 + \dfrac{40}{x}\cdot y = 60 + \dfrac{30}{x}\cdot y \quad$ gives $y = 2x$.

Answer: $L = 120$ steps.

116. We have to find the solution of the following system where x, y, z are natural numbers:

$$\left\{\begin{array}{l} x + y + z = 20 \\ 2x + 0.5y + 0.25z = 20 \end{array}\right..$$

Answer: 8 men, 4 women, 8 children.

117. $\sqrt{1995}\cdot x^{\log_{1995} x} = x^2 \implies x^{\log_x\sqrt{1995}}\cdot x^{\frac{1}{\log_x 1995}} = x^2 \implies \dfrac{1}{2}\log_x 1995 + \dfrac{1}{\log_x 1995} = 2$.

The solution of this equation is $x = 1995^{2\pm\sqrt{2}}$. Thus $x_1\cdot x_2 = 1995^{2+\sqrt{2}}\cdot 1995^{2-\sqrt{2}} = 1995^4$. The last three digits of this product are 625.

118. Let D be the distance between two villages and t be the time in hours for a friend from village A to reach the point of meeting. Note that the sum of the distances covered by each friend is equal to the distance between the villages. The following equation describes this situation: $\quad \dfrac{D}{4.5}t + \dfrac{D(t-1.5)}{3.6} = D$. The solution for t is

$t = 2\dfrac{5}{6}$ h or 2h 50 min. So the friends met at 11:28 AM.

119. Let x be the speed of the car from A and y be the speed of the car from B. Then the equation $2x + \dfrac{9}{8}y = 210$ reflects the fact that the sum of the distances covered by each car until their meeting is equal to the distance between villages. The second fact is that each car spent the same time from the point of departure to the point of meeting. So the second equation for x and y is

$$\frac{210 - 2x}{x} = \frac{210 - \dfrac{9}{8}y}{y}.$$ The solution of the system of equations is

$x = 60\,\dfrac{km}{h}, \quad y = 80\,\dfrac{km}{h}.$

120. $(0,0)$, $(2,2)$

121. Let $a = x + y$ and $b = \sqrt{xy}$. Then the system for the new variables is

$$\begin{cases} a^2 - 3b^2 = 61 \\ a - b = 7. \end{cases}$$ The solution is $a = 8$, $b = 1$ or $a = 13$, $b = 6$.

To find the initial variables we have to solve the following systems:

$$\begin{cases} x + y = 8 \\ \sqrt{xy} = 1 \end{cases} \quad \text{or} \quad \begin{cases} x + y = 13 \\ \sqrt{xy} = 6. \end{cases}$$

Answer: $(9,4)$, $(4,9)$, $\left(4 + \sqrt{15}, 4 - \sqrt{15}\right)$, $\left(4 - \sqrt{15}, 4 + \sqrt{15}\right)$

122. If the first number is x and the second is y, then the two statements can be interpreted by the system of equations

$$\begin{cases} x + y = 1244 \\ 10x + 3 = \dfrac{y}{10} - 0.2, \end{cases}$$ the solution of which is $x = 12$, $y = 1232$.

123. Let D be the discriminant of the given equation. Then the difference of the roots is $\dfrac{2\sqrt{D}}{4} = 1$ or $\sqrt{k^2 - 6k - 23} = 2$.

Answer: $k = 9$ or $k = -3$.

124. $k = \dfrac{2}{3}$.

125. Let x be the number of hundreds and y be the number of tens. Then the situation can be described by the equation $(100x + 10y + 3)3 = 300 + 10x + y - 1$. Obviously $x = 1$. Then $y = 0$. The original number is 103.

126. $(x + 1)(x + 2)(x + 3)(x + 4) = 24 \Leftrightarrow [(x + 1)(x + 4)][(x + 2)(x + 3)] = 24 \Leftrightarrow$ $(x^2 + 5x + 4)(x^2 + 5x + 6) = 24$. Introduce the new variable $y = x^2 + 5x + 5$. Then $(y - 1)(y + 1) = 24$; $y^2 - 1 = 24$; $y = \pm 5$. To find x we have to solve two equations: $x^2 + 5x + 5 = 5$ or $x^2 + 5x + 5 = -5$. The first equation has the solutions $x = 0$ and $x = -5$.

The second gives the complex solutions $x = \dfrac{-5 \pm i\sqrt{15}}{2}$.

127. $x = \dfrac{-5a \pm \sqrt{5a^2 \pm 4\sqrt{a^4 + b^4}}}{2}$. (**Hint:** Multiply the first factor by the fourth, and the second factor by the third. Then substitute $y = x^2 + 5ax + 5a^2$. See #126.)

128. Let D be the distance from A to B , x be the speed of the courier from A, and y be the speed of the courier from B. The first equation reflects the equality of the time each courier spent until their first meeting: $\dfrac{D-12}{x} = \dfrac{12}{y}$. The second equation we obtain by computing the time the first courier spent between meetings: $\dfrac{12}{x} + \dfrac{D-6}{x} = 6$. Analogously for the second carrier, we obtain the third equation: $\dfrac{D-12}{y} + \dfrac{6}{y} = 6$.

Answer: $D = 30$ km, $x = 6 \dfrac{km}{h}$, $y = 4 \dfrac{km}{h}$.

129. If x, y, z, and u are the time needed for each tap to fill the pool working alone, then we have the system of equations describing what part of the pool they can fill in one hour:

$$\begin{cases} \dfrac{1}{x} + \dfrac{1}{y} + \dfrac{1}{z} + \dfrac{1}{u} = \dfrac{1}{4}, \\ \dfrac{1}{x} + \dfrac{1}{y} + \dfrac{1}{z} = \dfrac{1}{5}, \\ \dfrac{1}{y} + \dfrac{1}{z} + \dfrac{1}{u} = \dfrac{1}{6}. \end{cases}$$

It is easy to find that the first and the fourth taps in one hour will fill $\dfrac{1}{x} + \dfrac{1}{u} = \dfrac{2}{15}$ part of the pool. Thus working together they need 7.5 hours to complete this job.

130. After dividing the left and right sides of the equation by 2, we have $\dfrac{x^2 + y^2}{2xy} + \dfrac{2xy}{x^2 + y^2} = 2$. Then $\dfrac{x^2 + y^2}{2xy} = 1 \Rightarrow (x - y)^2 = 0$ and $x = y \neq 0$.

(Note that $a + \dfrac{1}{a} = 2$ only if $a = 1$.)

131. 2.

132. (0,0), (1,0), (-1,0), (2,6), (2,-6), (3,12), (3,-12).

133. $(0,2)$, $(2,0)$, $\left(-1+\sqrt{5}, -1+\sqrt{5}\right)$, $(-1-\sqrt{5}, -1-\sqrt{5})$.
(**Hint:** Subtract the first equation from the second and consider two cases: $x = y$ and $x \neq y$.)

134. Excluding the variable y we have $(x - 1)^2 + z^2 = 0$. The only solution is $x = 1$, $u = 1$, $z = 0$.

45

135. At the end of this year the athlete had a weight of $0.75 \cdot 1.2 \cdot 0.9 \cdot 1.2x = 0.972x$ where x is his initial weight. So he lost 2.8% of the weight he had at the beginning of the year.

136. Denote by x, y and z the number of problems the student solved correctly, solved incorrectly and did not try to solve. Then we have the system of equations

$$\left\{ \begin{array}{l} 8x - 5y + 0 \cdot z = 13 \\ x + y + z = 20 \end{array} \right.$$

Only one natural number solution satisfies both equations: $x = 6$, $y = 7$, $z = 7$. Answer: 13.

137. Suppose that together with the flock of m white geese one additional gray goose is flying all time. Then at each lake exactly half of all geese are landing. $\dfrac{m}{2} + \dfrac{1}{2} = \dfrac{m+1}{2}$. After 7 lakes the number of geese decreased $2^7 = 128$ times and only the gray goose remains. So at the beginning there were 127 white and 1 gray goose.

Alternative solution: If x is the initial number of geese, then the numbers of geese which are landing at each lake are the terms of a geometric sequence:

$a_1 = \dfrac{1}{2}(x+1)$, $a_2 = \dfrac{1}{4}(x+1)$, $a_3 = \dfrac{1}{8}(x+1)$, ..., $a_7 = \dfrac{1}{128}(x+1)$. The sum of these terms must be x. The solution of this equation is $x = 127$.

138. $x^2 + 8x + 1 = 0$. (**Hint:** Decide what has to be the another root and then use Viete's Theorem: The solutions x_1 and x_2 of equation $x^2 + bx + c = 0$ must satisfy two conditions: $x_1 \cdot x_2 = c$ and $x_1 + x_2 = -b$.)

139. To find the interval of the speed of travel we have to solve the double inequality $3 < \dfrac{10}{x+1} + \dfrac{6}{x-1} < 4$ where x is the speed of the boat in still water and $x > 1$ (otherwise the boat would be unable to move against the stream.)

Answer: $4 \le x \le \dfrac{8 + \sqrt{61}}{3}$, x in miles per hour.

140. Let x and y be the time in days for each worker to do this job working alone. Then $\dfrac{1}{x}$ and $\dfrac{1}{y}$ are the portion of the job each of them does in one day working alone. So the situation can be described by the following system of equations:

$$\left\{ \begin{array}{l} \dfrac{1}{x} + \dfrac{1}{y} = \dfrac{1}{12} \\ \dfrac{1}{2} \div \dfrac{1}{x} + \dfrac{1}{2} \div \dfrac{1}{y} = 25 \end{array} \right.$$

Answer: 30 days and 20 days.

141.

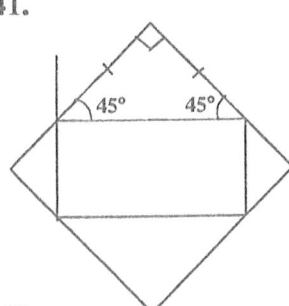

Note that each triangle formed by the bisectors of the exterior angles of the rectangle is a right isosceles triangle.

142.

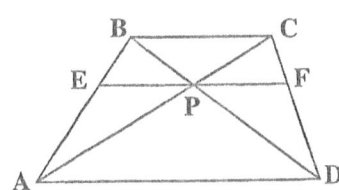

$\Delta BPC \sim \Delta DPA \Rightarrow \dfrac{AP}{PC} = \dfrac{AD}{BC};$ $\Delta BAC \sim \Delta EAP \Rightarrow \dfrac{EP}{BC} = \dfrac{AP}{AC};$

$\Delta ACD \sim \Delta PCF \Rightarrow \dfrac{PF}{AD} = \dfrac{AP}{AC};$ After dividing the second proportion

by the third we have $\dfrac{EP}{BC} \cdot \dfrac{AD}{PF} = \dfrac{AP}{PC}$ $\Rightarrow \dfrac{EP}{PF} = \dfrac{AP}{PC} \cdot \dfrac{BC}{AD} = 1 \Rightarrow$

$EP = PF$.

143.

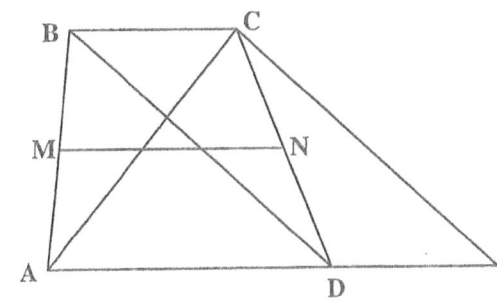

Draw $CF \parallel BD$. $\angle ACF = 90°$. So ΔACF is right. Then $AF^2 = AC^2 + CF^2$, but $CF = BD$, so $AF^2 = AC^2 + BD^2 = 169$. $AF = 13$, $AF = AD + BC$, $AD + BC = 13$.

$MN = \dfrac{AD + BC}{2} = 6.5$.

144.

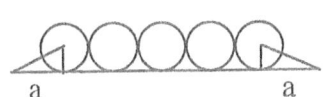

Note that $a = 2.5\sqrt{3}$

Answer: $20 + 2.5\sqrt{3}$.

145.   Area = ab

146.

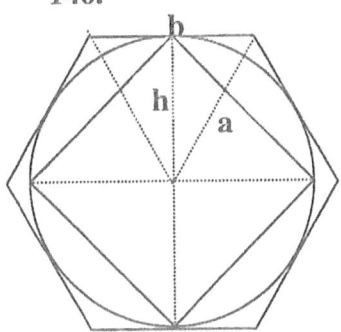

Notice: $h = \dfrac{a}{\sqrt{2}}, b = \dfrac{2h}{\sqrt{3}}$.

Answer: Area $= a^2\sqrt{3}$.

147.

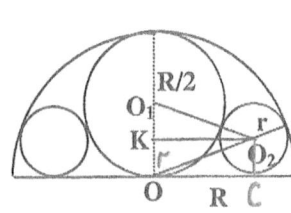

$$OO_1 = \frac{R}{2} + r, \quad OK = \frac{R}{2} - r; \quad OC = \sqrt{\left(\frac{R}{2} + r\right)^2 - \left(\frac{R}{2} - r\right)^2} = \sqrt{2Rr}.$$

$$OO_2 = R - r = \sqrt{OC^2 + r^2}; \quad 2Rr + r^2 = R^2 - 2Rr + r^2 \Rightarrow r = \frac{R}{4}.$$

$$Area = \frac{\pi R^2}{2} - \pi\left(\frac{R}{2}\right)^2 - 2\pi\left(\frac{R}{4}\right)^2 = \frac{\pi R^2}{8}.$$

148. Draw a diameter of the circle through the given point. Then draw a chord through this point perpendicular to the diameter.

149. Area $= 4\sqrt{3} \, in^2$. (**Hint:** A median divides a triangle into two triangles of equal area.)

150.

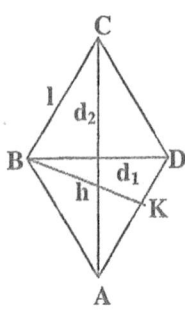

"The side of a rhombus is the geometric mean of the diagonals" means that $l^2 = d_1 \cdot d_2$. The area of rhombus $A = \frac{1}{2} d_1 \cdot d_2 = \frac{1}{2} l^2$.

On the another hand $A = h \cdot l$. So $\frac{1}{2} l^2 = l \cdot h$ and $h = \frac{1}{2} l$. Thus in triangle ABK $\angle A = 30°$. Then $\angle B = 150°$.

151.

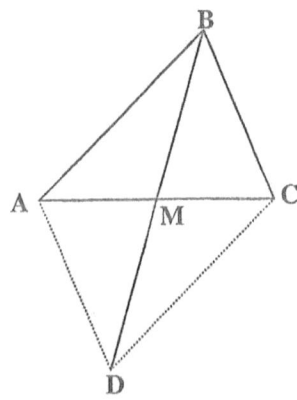

Let $AB = 12cm$, $BC = 11cm$, $AC = 13cm$. Draw $CD \parallel AB$ and $AD \parallel BC$. Then ABCD is a parallelogram and $AC^2 + BD^2 = 2AB^2 + 2BC^2$.

So $BM = \frac{1}{2}\sqrt{2 \cdot 11^2 + 2 \cdot 12^2 - 13^2} = 9.5 \, cm$.

(By the way we proved the formula for the length of a median

$$m_c = \frac{1}{2}\sqrt{2a^2 + 2b^2 - c^2}$$

where a, b, c are the sides of a triangle and m_c is the median to side c .)

152.

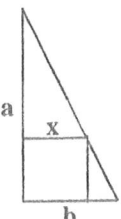

$h = l\sin\alpha;$ $\quad AN = l\cos\alpha + x;$ $\quad BC = l\cos\alpha - x$. Area $= \dfrac{1}{2}h(BC + AD);$

Area $= \dfrac{1}{2}l\sin\alpha(l\cos\alpha + x + l\cos\alpha - x) = \dfrac{l^2 \cdot 2\sin\alpha\cos\alpha}{2} = \dfrac{l^2\sin 2\alpha}{2}.$

153.

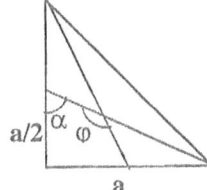

Using similarity of triangles we have $\dfrac{a - x}{x} = \dfrac{x}{b - x};$ $\quad x = \dfrac{ab}{a + b}.$

The perimeter $= \dfrac{4ab}{a + b}.$

154. Answer: $\sqrt{10}$cm. (See #151.)

155.

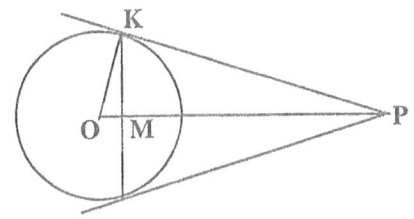

$\tan\alpha = 2;$ $\quad \varphi = 270° - 2\tan^{-1}2 \approx 143.13°$.

156.

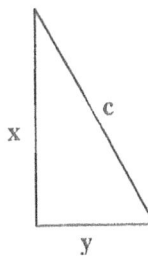

Examine ΔMKP : $MP = \sqrt{KP^2 - PM^2} = 9.6$cm.
Examine ΔOKP: $\angle OKP = 90°$; $KM^2 = OM\cdot MP \Rightarrow OM = 5.4$cm.
$R = OK = \sqrt{7.2^2 - 5.4^2} = 9$cm.

157.

Let A be the area of the triangle. Let us create a function of one variable which represents the area. $A(x,y) = \dfrac{1}{2}xy$ and $y = \sqrt{c^2 - x^2}$.

So $A(x) = \dfrac{1}{2}x\sqrt{c^2 - x^2} = \dfrac{1}{2}\sqrt{c^2x^2 - x^4}$. $A'(x) = \dfrac{c^2 \cdot 2x - 4x^3}{2 \cdot 2\sqrt{c^2x^2 - x^4}} = \dfrac{c^2x - 2x^3}{2\sqrt{c^2x^2 - x^4}}.$

In order to get the maximum area we have to solve the equation $A'(x) = 0.$

The solution of this equation is $x = \dfrac{c}{\sqrt{2}}$. Thus the triangle must be isosceles : $a = b$.

(Note that this problem relates to maximizing the area of a rectangle with a given perimeter, and might be solved without using calculus.)

49

158.

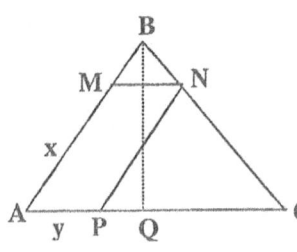

Area $= xy\sin\angle A$ where x and y are the sides of the parallelogram. Note that $\sin\angle A = \dfrac{BQ}{AQ} = \dfrac{12}{15} = \dfrac{4}{5}$. So $A(x,y) = \dfrac{4}{5}xy$.

$\Delta MBR \sim \Delta ABQ \Rightarrow \dfrac{15-x}{15} = \dfrac{y}{2\cdot 9} \Rightarrow y = \dfrac{90-6x}{5}$. So $A(x) = \dfrac{360-24x^2}{25}$.

$A'(x) = \dfrac{360-48x}{25}$. $A'(x)=0$ at $x=7.5$in and $y=9$in.

159.

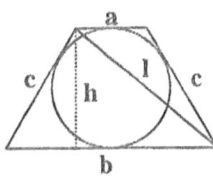

Note that $a+b=2c$; $c = \dfrac{a+b}{2}$; $h = \sqrt{\left(\dfrac{a+b}{2}\right)^2 - \left(\dfrac{a-b}{2}\right)^2} = \sqrt{ab}$.

Then the area of the circle $= \pi r^2 = \pi\left(\dfrac{\sqrt{ab}}{2}\right)^2 = \dfrac{\pi ab}{4}$.

The length of the diagonal $l = \sqrt{\dfrac{a^2+3ab+b^2}{2}}$.

160.

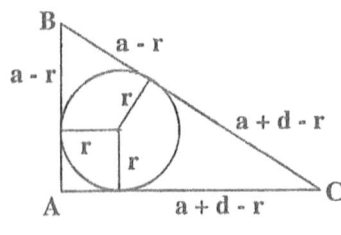

The equation of a circle is $x^2+y^2=R^2$. The coordinates of a vertex of the square are $x = \dfrac{a}{2}, y=a$. So $\dfrac{a^2}{4}+a^2 = R^2$ and

$$a = \dfrac{2R\sqrt{5}}{5}.$$

161.

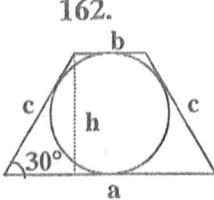

Let $AB=a$, $AC=a+d$, $BC=a+2d$. Then $a+d-r+a-r=a+2d \Rightarrow a = d+2r$. By the Pythagorean Theorem $a^2+(a+d)^2=(a+2d)^2$ e.g. $a^2-2ad-3d^2=0$. After substituting $a=d+2r$ we have $4r^2=4d^2 \Rightarrow r=d$.

162.

Note that $a+b=2c$. The area of the trapezoid $= \dfrac{1}{2}h(a+b) = \dfrac{1}{2}\cdot\dfrac{c}{2}\cdot 2c = 2 \Rightarrow c=2$.

$b = a+2\cdot\dfrac{c\sqrt{3}}{2} = a+2\sqrt{3}$; $2a+2\sqrt{3}=4$; $a=2-\sqrt{3}$ and $b=2+\sqrt{3}$.

163.

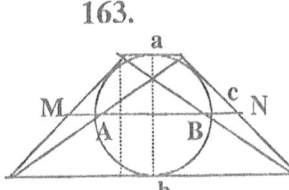

$MA = BN = \dfrac{a}{2}$; $a + b = 2c$; $AB = \dfrac{3a}{2}$; $MN = \dfrac{5a}{2}$; MN is a midsegment

of the trapezoid, i.e. $MN = \dfrac{a+b}{2}$; $b = 4a$. Thus the perimeter of the

trapezoid is equal to 10a. The radius of the circle $r = \dfrac{1}{2}\sqrt{\left(\dfrac{5a}{2}\right)^2 - \left(\dfrac{3a}{2}\right)^2} = a$.

$\dfrac{\text{Perimeter}}{\text{Circumference}} = \dfrac{10a}{2\pi a} = \dfrac{5}{\pi}$.

164.

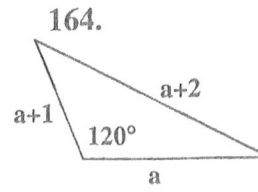

By the Cosine Theorem $(a + 2)^2 = a^2 + (a + 1)^2 - 2a(a + 1)\cos 120°$.

The solution of this equation is $a = \dfrac{3}{2}$.

Answer: $\dfrac{3}{2}, \dfrac{5}{2}, \dfrac{7}{2}$.

165.

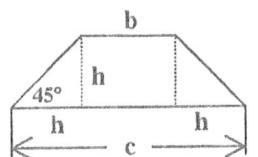

$\text{Area} = \dfrac{1}{2}h(b+c)$. We will represent the area as a function of one

variable h. $h + c = a$; $c = b + 2h$; $b + 3h = a$; $b = a - 3h$.

Then $A(h) = h(a - 2h) = -2h^2 + ah$. This function reaches its

maximum at $h = \dfrac{a}{4}$. Thus the trapezoid of maximum area has

sides $\dfrac{1}{4}a, \dfrac{3}{4}a, \dfrac{1}{4}a\sqrt{2}, \dfrac{1}{4}a\sqrt{2}$.

166.

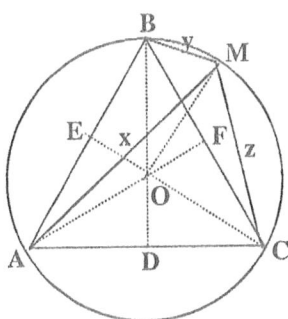

Let the radius of the circumscribed circle be R, $\angle BOM = 2\alpha$, $\angle MOC = 2\beta$,

$\angle AOM = 2\gamma$ and $MA = x$, $MB = y$, $MC = z$. Let us find how x, y, z

depend on R and α: $y = 2R\sin\alpha$; $2\beta = 120° - 2\alpha$; $\beta = 60° - \alpha$;

$z = 2R\sin(60° - \alpha)$; $y + z = 2R(\sin\alpha + \sin(60° - \alpha)) = 4R\sin 30°\cos(\alpha - 30°) = 2R\cos(\alpha - 30°)$.

On the another hand $2\gamma = 120° + 2\alpha$; $\gamma = 60° + \alpha$; $x = 2R\sin(60° + \alpha) = 2R\cos(30° - \alpha) = 2R\cos(\alpha - 30°)$, so $x = y + z$.

Let $BF = FC = a$ and $AF = b$. Then by the Sine Theorem, in

$\triangle AFB$ $\dfrac{a}{b} = \dfrac{\sin 3\alpha}{\sin(90 - \alpha)}$ and in $\triangle ACF$ $\dfrac{a}{b} = \dfrac{\sin\alpha}{\sin(90 - 3\alpha)}$.

167.

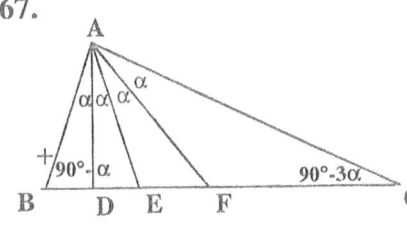

So we have to solve the equation $\dfrac{\sin 3\alpha}{\cos\alpha} = \dfrac{\sin\alpha}{\cos 3\alpha}$;

$\sin\alpha \cdot \cos\alpha = \sin 3\alpha \cdot \cos 3\alpha$; $\sin 2\alpha = \sin 6\alpha$; $2\cos 4\alpha \cdot \sin 2\alpha = 0$;

$\cos 4\alpha = 0$; $\Rightarrow 4\alpha = 90$; $\sin 2\alpha = 0 \Rightarrow \alpha = 90$ (impossible).

168.

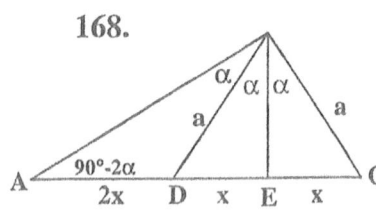

Note that in $\triangle EBC$ $a = \dfrac{x}{\sin\alpha}$. By the Sine Theorem for $\triangle ABD$,

we have $\dfrac{2x}{\sin\alpha} = \dfrac{x}{\sin\alpha \cdot \sin(90 - 2\alpha)}$ $\Rightarrow \sin(90° - 2\alpha) = \dfrac{1}{2}$ $\Rightarrow$

$\alpha = 30°$ and $\angle ABC = 90°$.

169.

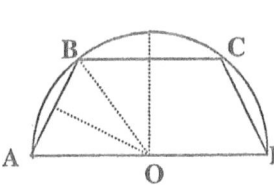

Let $AB = a$, $BC = b$, $\angle BAD = x$. Then $\dfrac{a}{2} = R\cos x$, $a = 2R\cos x$,

$\dfrac{b}{2} = R\cos(\pi - 2x)$, $b = -2R\cos 2x$. The perimeter $P = 2R + 2a + b$ as a function

of x is $P(x) = 2R + 4R\cos x - 2R\cos 2x$. $P'(x) = -4R\sin x + 4R\sin 2x$;

$P'(x) = 0$; $\sin 2x - \sin x = 0$; $\sin x(2\cos x - 1) = 0$. Then either $\sin x = 0$;

$x = \pi n$ (impossible) or $2\cos x - 1 = 0$, $x = \dfrac{\pi}{3}$.

170.

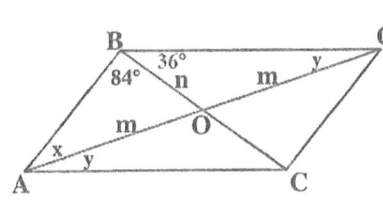

In the figure, by the Sine Theorem for $\triangle ABO$ and $\triangle BCO$ we have

$\dfrac{m}{\sin 84°} = \dfrac{n}{\sin x}$ and $\dfrac{m}{\sin 36°} = \dfrac{n}{\sin y}$. Therefore $\dfrac{\sin 86°}{\sin x} = \dfrac{\sin 36°}{\sin y} = \dfrac{\sin 36°}{\sin(60° - x)}$.

Now we have to solve the trigonometric equation

$\sin 36° \cdot \sin x = \sin 86° \cdot \sin(60° - x)$. Using the formula for the sine of the

difference of two angles and dividing the left and right sides of the

equation by $\cos x$, we have $\tan x = \dfrac{\sqrt{3}}{2\left(\dfrac{\sin 36°}{\sin 84°} + \dfrac{1}{2}\right)}$.

Answer: $x \approx 38.4°$, $y \approx 21.6°$.

171. $\tan(\alpha + \beta) = \dfrac{\tan\alpha + \tan\beta}{1 - \tan\alpha \cdot \tan\beta} = \dfrac{\dfrac{1}{2} + \dfrac{1}{3}}{1 - \dfrac{1}{2} \cdot \dfrac{1}{3}} = 1 \Rightarrow \alpha + \beta = 45°$.

172. Construct an angle A equal to the given angle α. Construct a perpendicular to the side AB_1 from an arbitrary point B_1 on this side. Construct B_1C_1 congruent to the given leg. Through point C_1 construct a line parallel to the side AB_1. Let C be the point of intersection of this line with the second side of angle A. Then from point C construct a perpendicular to the side AB_1. Triangle ABC is the desired triangle.

Given:

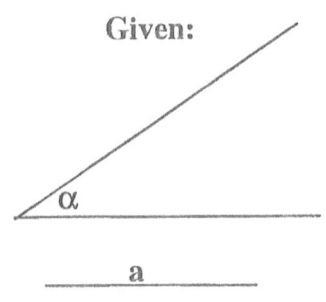

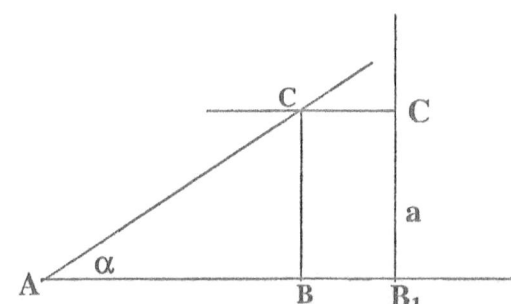

52

173. Through an arbitrary point on a line construct a perpendicular to this line. Let DB be a copy of the given altitude h. Divide the given altitude into three equal parts and construct $BO = \frac{2}{3}h$. Point O is the point of intersection of the medians of the triangle. Construct a circle with a radius of $\frac{2}{3}m$ and center O. Triangle ABC is the desired triangle.

Given:

h

m

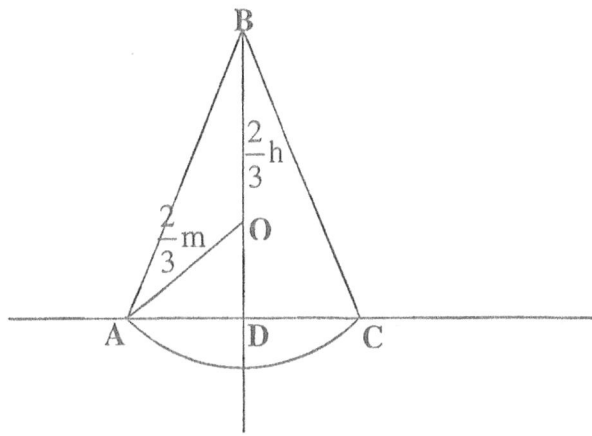

174. Copy segment a on a given line. Construct a perpendicular bisector to this segment. Construct a circle of radius R with center at the end of the segment. The point of intersection of this circle with the perpendicular bisector is the center of the desired circle. Note that the distance from the center of the desired circle to the given line must be less that the radius of the circle.

Given:

a

R

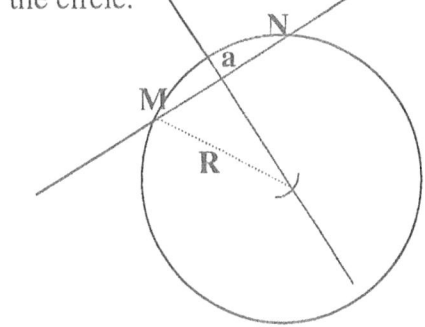

175. On an arbitrary line MN chose any points A and D_1 and construct copies of the given angles with these vertices. Copy segment AB = a and $C_1D_1 = c$. Through point C_1 construct a line parallel to MN. Construct an arc with radius b and center at point B. The quadrilateral ABCD is the desired quadrilateral.

Given:

a

b

c

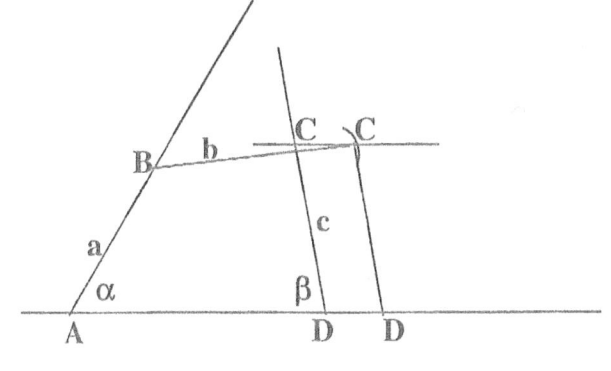

53

176. Let a and b be the bases and c and d be the nonparallel sides of the trapezoid. Construct a triangle with sides c, d, and $b - a$. Then construct a parallelogram with sides a and c. ABCD is the desired trapezoid.

Given:

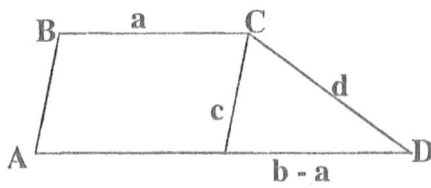

177. Hint: Construct any circle inscribed in the given angle. Then use the fact that all circles are similar.

178. Hint: Use the theorem about the point of intersection of the medians of a triangle.

179. To find the angle we choose a system of coordinates.

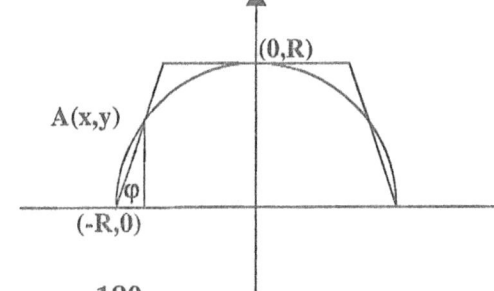

The ordinate of point A is $y = \dfrac{R}{2}$. Then the abscissa of this point satisfies the equation $x^2 + \left(\dfrac{R}{2}\right)^2 = R^2$; $x = -\dfrac{R\sqrt{3}}{2}$.

$$\tan\varphi = \frac{\Delta y}{\Delta x} = \frac{\dfrac{R}{2} - 0}{-\dfrac{R\sqrt{3}}{2} + R} = \frac{R}{R\left(2 - \sqrt{3}\right)} = 2 + \sqrt{3}; \qquad \varphi = 75°.$$

(Note that $\tan 75° = 2 + \sqrt{3}$ precisely.)

180.

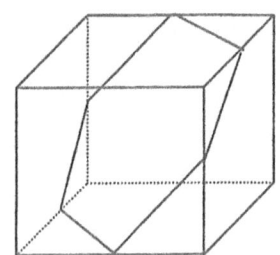

The section is a regular hexagon with a side of $\dfrac{a\sqrt{2}}{2}$.

The area of the hexagon is equal to $\dfrac{3a^2\sqrt{3}}{4}$ un^2.

181.

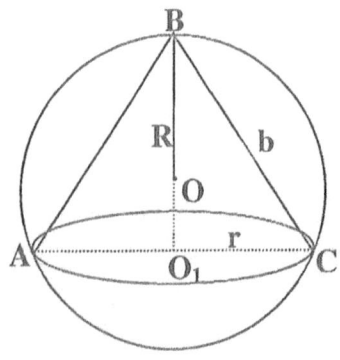

Let O be the center of the sphere and O_1 the center of the base of the inscribed cone. Then the radius of the sphere $R = \dfrac{2}{3}BO_1 = \dfrac{2b\sqrt{3}}{3 \cdot 2} = \dfrac{b\sqrt{3}}{3}$.

The surface area of the sphere $SA_{sph} = 4\pi R^2 = \dfrac{4\pi b^2}{3}$.

The surface area of the cone $SA_{cone} = \pi r^2 + \pi r b = \pi\left(\dfrac{b}{2}\right)^2 + \pi\left(\dfrac{b}{2}\right)b = \dfrac{3\pi b^2}{4}$.

Thus $\dfrac{SA_{cone}}{SA_{sph}} = \dfrac{9}{16}$.

182.

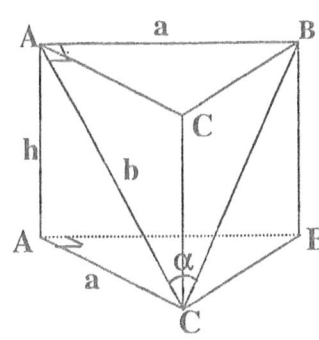

AC is the projection of AC_1 onto the plane (ABC), so $\angle BAC_1 = 90°$.

In $\triangle ABC_1$: $b = a\cot\alpha$. In $\triangle A_1AC_1$: $h = \sqrt{a^2\cot^2\alpha - a^2}$.

The volume of the prism is $V = S_{base}h = \dfrac{1}{2}a^2 \cdot a\sqrt{\cot^2\alpha - 1} = \dfrac{a^3\sqrt{\cos 2\alpha}}{2\sin\alpha}$.

183. $45°$.

184.

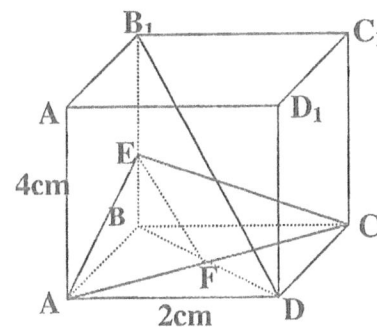

$B_1D = \sqrt{B_1B^2 + BD^2} = \sqrt{\left(2\sqrt{2}\right)^2 + 4^2} = 2\sqrt{6}$. Note that $B_1D \parallel EF$.

So $\triangle BEF \sim \triangle BB_1D \Rightarrow \dfrac{EF}{B_1D} = \dfrac{BF}{BD} = \dfrac{1}{2}$; $EF = \sqrt{6}$.

The area of the section $= \dfrac{1}{2}EF \cdot AC = \dfrac{1}{2}\sqrt{6} \cdot 2\sqrt{2} = 2\sqrt{3} \text{cm}^2$.

185.

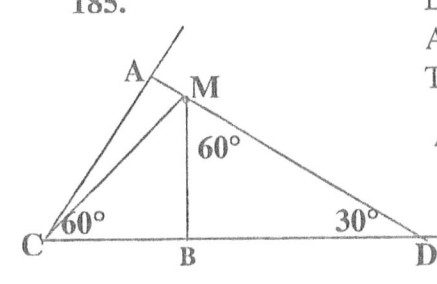

Let $AM = 2$cm, $BM = 11$cm. Extend AM; D is the intersection of line AM with side CB. Note that $\angle BMD = 60°$ ($BM \perp CD$ and $MD \perp CA$). Then $\angle MDB = 30°$. In $\triangle BMD$ $MD = 2BM = 22$ cm. In $\triangle CAD$ $AD = 24$cm; $CA = \dfrac{24}{\sqrt{3}} = 8\sqrt{3}$cm. In $\triangle CAM$ $CM = \sqrt{\left(8\sqrt{3}\right)^2 + 2^2} = 14$cm.

186. Hint: From a vertex of the trapezoid draw a line parallel to the diagonal of the trapezoid.

187. The lengths of the diagonals of such quadrilaterals are equal, because each of them is equal to double the length of the segment that joints the midpoints of the sides. Moreover, the diagonals of those quadrilaterals are parallel (apply the midsegment theorem). The area of a quadrilateral is equal to the product of the lengths of the diagonals by the sine of the angle between them (see (18)). So by the previous observations, the areas of the described quadrilaterals are equal.

188.

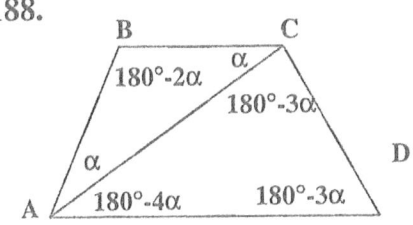

$180° - 4\alpha = \alpha \Rightarrow \alpha = 36°$.

$\angle A = \angle D = 72°$; $\angle B = \angle C = 108°$.

189.

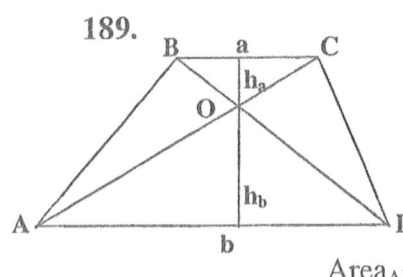

$$\Delta AOD \sim \Delta COB \Rightarrow \frac{h_a^2}{h_b^2} = \frac{S_1}{S_2} \text{ and } \frac{a^2}{b^2} = \frac{S_1}{S_2} \Rightarrow$$

$$h_a = h_b \sqrt{\frac{S_1}{S_2}} \text{ and } a = b\sqrt{\frac{S_1}{S_2}}.$$

$$\text{Area}_{ABCD} = \frac{1}{2}(h_a + h_b)(a+b) = \frac{1}{2}h_b\left(\sqrt{\frac{S_1}{S_2}}+1\right)\cdot b\left(\sqrt{\frac{S_1}{S_2}}+1\right) = \left(\sqrt{S_1}+\sqrt{S_2}\right)^2.$$

190.

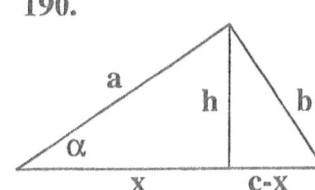

Given that $4abh = abc$. So $c = 4h$. On the another hand $h^2 = x(4h-x)$. Solving this equation for x we have $x = h\left(2\pm\sqrt{3}\right)$.

$$\tan\alpha = \frac{h}{x} = \frac{h}{h\left(2\pm\sqrt{3}\right)} = 2\mp\sqrt{3} \Rightarrow \alpha = 15° \text{ or } 75°.$$

191.

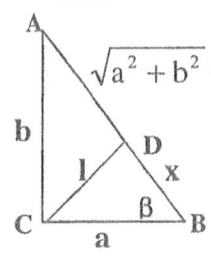

By properties of bisectors $\dfrac{a}{b} = \dfrac{x}{\sqrt{a^2+b^2}-x}$ whence $x = \dfrac{a\sqrt{a^2+b^2}}{a+b}$.

Using the Sine Theorem for triangle CBD we have

$$\frac{l}{\sin\beta} = \frac{x}{\sin 45°} \Rightarrow \frac{l}{\dfrac{b}{\sqrt{a^2+b^2}}} = \frac{x}{\dfrac{\sqrt{2}}{2}} \Rightarrow l = \frac{ab\sqrt{2}}{a+b}.$$

192.

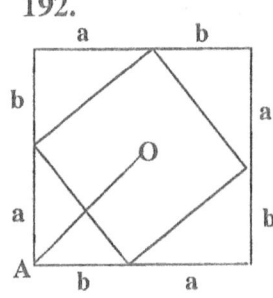

On each side of the square construct a triangle congruent to the given triangle (see the figure). The legs of these triangles form a square, the center of which matches with the center of the original square. The unknown distance is equal to half of the diagonal of the new square.

Answer: $AO = \dfrac{\sqrt{2}}{2}(a+b)$.

193. Yes. Consider the example where the first triangle is an equilateral triangle with a side of 0.5 centimeters , and the second triangle is an isosceles triangle with a side of 200 meters and an altitude of 10^{-7}m.

194.

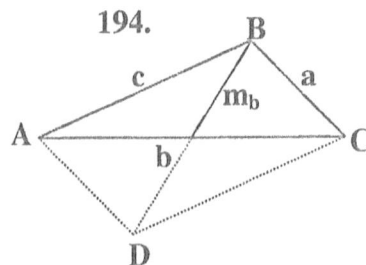

Construct CD parallel to AB and AD parallel to BC. ABCD is a parallelogram . Then $AC^2 + BD^2 = 2AB^2 + 2BC^2$.

So $b^2 + (2m_b)^2 = 2a^2 + 2c^2 \Rightarrow m_b = \dfrac{1}{2}\sqrt{2a^2 + 2c^2 - b^2}$.

195. $\cos 15° = \sqrt{\dfrac{1+\cos 30°}{2}}$; $\quad \sin 15° = \sqrt{\dfrac{1-\cos 30°}{2}}$. $\quad \dfrac{\log 2\cos 15°}{\log 2\sin 15°} = \dfrac{\log 2\dfrac{\sqrt{2+\sqrt 3}}{2}}{\log 2\dfrac{\sqrt{2-\sqrt 3}}{2}} = \dfrac{\log\sqrt{2+\sqrt 3}}{-\log\sqrt{2+\sqrt 3}} = -1.$

196. $\cos 20°\cdot\cos 40°\cdot\cos 80° = \dfrac{2\sin 20°\cdot\cos 20°\cdot\cos 40°\cdot\cos 80°}{2\sin 20°} = \dfrac{2\sin 40°\cdot\cos 40°\cdot\cos 80°}{4\sin 20°} = \dfrac{2\sin 80°\cdot\cos 80°}{8\sin 20°}$

$\dfrac{\sin 160°}{8\sin 20°} = \dfrac{\sin 20°}{8\sin 20°} = \dfrac{1}{8}.$

197. Note that $\sin 90° = 1$, $\sin 70° = \cos 20°$, and $\sin 50° = \cos 40°$. (**Hint:** Multiply the numerator and denominator of the expression by $2\cos 10°$ and repeatedly use the formula for the sine of a double angle.)

198. $\sin 47° + \sin 61° - \sin 11° - \sin 25° = (\sin 47° - \sin 11°) + (\sin 61° - \sin 25°) = $
$2\cos 29°\cdot\sin 18° + 2\cos 43°\cdot\sin 18° = 2\sin 18°(\cos 29° + \cos 43°) = 2\sin 18°\cdot 2\cos 36°\cdot\cos 7° = $
$\dfrac{2\sin 18°\cdot\cos 18°\cdot 2\cos 36°\cdot\cos 7°}{\cos 18°} = \dfrac{2\sin 36°\cdot\cos 36°\cdot\cos 7°}{\cos 18°} = \dfrac{\sin 72°\cdot\cos 7°}{\cos 18°} = \cos 7°.$

199. Note that $\quad x \neq \dfrac{\pi}{2} + \pi n$ where $n = 0, \pm 1, \pm 2, \ldots$

$\cos x = \dfrac{\sin x}{\cos x}\cdot 2\cos^2 x \Leftrightarrow \cos x = 2\sin x\cdot\cos x$. Since $\cos x \neq 0$, $\sin x = \dfrac{1}{2}$.

Thus $x = (-1)^k\dfrac{\pi}{6} + \pi k$ where $k = 0, \pm 1, \pm 2, \pm 3, \ldots$

200. $\cos x\cdot\tan^2\dfrac{x}{2} = -\dfrac{3}{2}$, $x \neq \pi + 2\pi k$, $k = 0, \pm 1, \pm 2, \ldots$; $\qquad \cos x\cdot\dfrac{1-\cos x}{1+\cos x} = -\dfrac{3}{2}$.
$2\cos x(1-\cos x) + 3(1+\cos x) = 0$; $\quad 2\cos^2 x - 5\cos x - 3 = 0$; $\quad \cos x = -0.5$ or $\cos x = 3 > 1$.

Thus $\quad x = \pm\dfrac{2}{3}\pi + 2\pi n$, $n = 0, \pm 1, \pm 2, \pm 3, \ldots$

201.
$\sin^4 x - \cos^4 x = \cos\left(\dfrac{3}{2}\pi - x\right) \Leftrightarrow \sin^2 x - \cos^2 x = -\sin x \Leftrightarrow 2\sin^2 x + \sin x - 1 = 0.$

Answer: $\quad x = (-1)^k\dfrac{\pi}{6} + 2\pi k$, $k = 0, \pm 1, \pm 2, \pm 3, \ldots$ or $x = \dfrac{3}{2}\pi + 2\pi n$, $n = 0, \pm 1, \pm 2, \pm 3, \ldots$

202. $\cos x \cdot \cos 3x = \cos 5x \cdot \cos 7x \Leftrightarrow \cos 4x + \cos 2x = \cos 12x + \cos 2x \Leftrightarrow \cos 4x - \cos 12x = 0 \Leftrightarrow$

$2\sin 8x \cdot \sin 4x = 0$.

$$\sin 8x = 0 \qquad \text{or} \qquad \sin 4x = 0$$

$$x = \frac{\pi}{8}k, \quad k = 0, \pm 1, \pm 2, \ldots \qquad x = \frac{\pi}{4}n, \quad n = 0, \pm 1, \pm 2, \ldots$$

Thus the minimum solution is $\frac{\pi}{8}$.

203. $\dfrac{\cos x}{\sin x} - \dfrac{\sin x}{\cos x} = \sin x + \cos x; \qquad \dfrac{\cos^2 x - \sin^2 x}{\cos x \cdot \sin x} = \sin x + \cos x$. Now we have two cases:

$$\sin x + \cos x = 0 \qquad \text{or} \qquad \sin x + \cos x \neq 0$$

$$\tan x = -1 \qquad \qquad \dfrac{\cos x - \sin x}{\sin x \cdot \cos x} = 1 \Rightarrow \cos x - \sin x = \sin x \cdot \cos x \Rightarrow$$

$$x = -\frac{\pi}{4} + \pi n, \quad n = 0, \pm 1, \pm 2, \ldots \qquad \cos^2 x - 2\sin x \cdot \cos x + \sin^2 x = \sin^2 x \cdot \cos^2 x \Rightarrow$$

$$1 - \sin 2x = \frac{1}{4}\sin^2 2x \Leftrightarrow \sin^2 2x + 4\sin 2x - 4 = 0$$

$$\sin 2x = -2 \pm 2\sqrt{2}; \qquad x = \frac{1}{2}\left[(-1)^k \arcsin 2\left(\sqrt{2} - 1\right) + \pi k\right], \quad k = 0, \pm 1, \pm 2, \ldots$$

204. $\begin{cases} \tan x + \tan y = 2 \\ 2\cos x \cdot \cos y = 1 \end{cases} \Leftrightarrow \begin{cases} \dfrac{\sin(x + y)}{\cos x \cdot \cos y} = 2 \\ 2\cos x \cdot \cos y = 1 \end{cases}$.

Multiplying the left and right sides of the first and second equations we have

$\begin{cases} \sin(x + y) = 1 \quad \Rightarrow \quad x + y = \dfrac{\pi}{2} + 2\pi m. \text{ (Note that } \cos(x + y) = 0.) \\ 2\cos x \cdot \cos y = 1 \quad \Leftrightarrow \quad \cos(x + y) + \cos(x - y) = 1 \quad \Rightarrow \quad \cos(x - y) = 1 \Rightarrow x - y = 2\pi k. \end{cases}$

Now we have the system:

$\begin{cases} x + y = \dfrac{\pi}{2} + 2\pi m \\ x - y = 2\pi k. \end{cases}$ Answer: $x = \dfrac{\pi}{4} + (n + k)\pi;$; $y = \dfrac{\pi}{4} + (n - k)\pi$. $n = 0, \pm 1, \pm 2, \ldots,$ $k = 0, \pm 1, \pm 2, \ldots$

205. $\cos x + \sqrt{3}\sin x \leq 2 \Leftrightarrow \dfrac{1}{2}\cos x + \dfrac{\sqrt{3}}{2}\sin x \leq 1 \Leftrightarrow \sin(x + 30°) \leq 1$.

206. $\dfrac{1}{4}$. **(Hint:** Multiply and divide the given expression by $2\sin\dfrac{2\pi}{5}$. Then use the formula for the sine of a double angle.)

207. If $\tan\alpha = \dfrac{1}{7}$, then $\cos^2\alpha = \dfrac{49}{50}$ and $\sin^2\alpha = \dfrac{1}{50}$. $\cos 2\alpha = \cos^2\alpha - \sin^2\alpha = \dfrac{24}{25}$.

If $\tan\beta = \dfrac{1}{3}$, then $\cos^2\beta = \dfrac{9}{10}$, $\sin^2\beta = \dfrac{1}{10}$, $\cos 2\beta = \dfrac{4}{5}$, $\sin 2\beta = \dfrac{3}{5}$,

$\sin 4\beta = 2\sin 2\beta \cdot \cos 2\beta = \dfrac{24}{25}$.

208.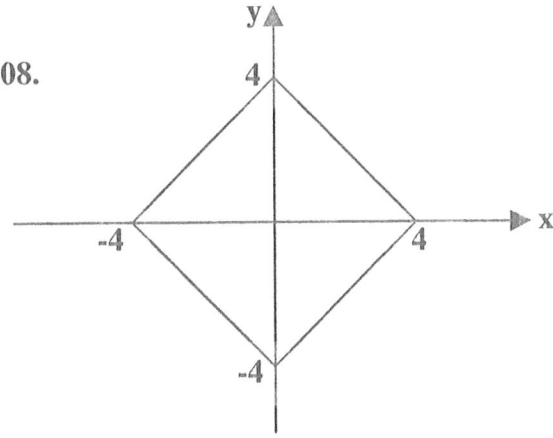

209. If $f(x)$ is odd, then $f(x) + f(-x) = 0$. If $f(x)$ is even, then $f(x) + f(-x) = 2f(x)$.

In our case $f(x) + f(-x) = \log(\sqrt{9\tan^2 x + 1} + 3\tan x) + \log(\sqrt{9\tan^2(-x) + 1} + 3\tan(-x)) =$

$\log\left[\left(\sqrt{9\tan^2 x + 1} + 3\tan x\right)\left(\sqrt{9\tan^2 x + 1} - 3\tan x\right)\right] = \log\left(9\tan^2 x + 1 - 9\tan^2 x\right) = \log 1 = 0$.

Thus $f(x)$ is an odd function.

210. $(-5, 0)$.

211. Let x and y be the unknown addends of 10. Then we have two conditions:

$x + y = 10$ and $f(x,y) = \dfrac{1}{2}x^2 + y^3$ has to be a minimum. Substituting for x we have

a function of one variable to minimize : $f(y) = \dfrac{1}{2}(10 - y)^2 + y^3$. $f'(y) = y - 10 + 3y^2$;

$f'(y) = 0$ at $y = -2$ and $y = \dfrac{5}{3}$. The negative solution is not acceptable, so $x = \dfrac{25}{3}$, $y = \dfrac{5}{3}$.

212. $(0,3) \cup (3,6)$

213. To find the domain we have to solve two inequalities: $x + 0.5 > 0$; $x \neq 1$ and

$\dfrac{x^2 + 2x - 3}{4x^2 + 4x - 3} > 0$. The second inequality has the solution $(-1.5, 0.5) \cup (1, \infty)$. The

intersection of this set with the set of numbers $x > -0.5$ gives the answer :

$(-0.5, 0.5) \cup (1, \infty)$.

214. The domain of the function is restricted by the conditions :

$4 - x^2 \geq 0$, $-1 \leq \dfrac{1}{x} \leq 1$ and $x \equiv 2$. Answer: $[-2, -1] \cup [1, 2)$.

215. $a = 1$. (**Hint:** Check that $x_1^2 + x_2^2 = a^2 - 2a + 4$. The graph of this function of a is

a parabola with the minimum value at the point $a = \dfrac{-(-2)}{2} = 1$.)

216. a) Yes. Let us look for such a polynomial in the form $P(x) = ax^2 + bx + c$.
Substituting $x = 0$, $x = 1$ and $x = 2$ we have a system of equations with respect to the variables a, b and c:

$$\begin{cases} a + b + c = 85 \\ 4a + 2b + c = 1985 \\ c = 19 \end{cases}$$, the solution of which is $a = 917$, $b = -851$, $c = 19$.

So we found the polynomial $P(x) = 917x^2 - 851x + 19$.

b) No. Note that if a polynomial $P(x)$ has integer coefficients, then $P(a) - P(b)$ must be divisible by $a - b$. In our case $P(19) - P(1) = 66$ is not divisible by $19 - 1 = 18$.

217. If the equation $ax^2 + bx + c = 0$ does not have real roots, the graph of the function $f(x) = ax^2 + bx + c$ is a parabola located either above or below the x-axis. Note that $f(1) = a + b + c < 0$, so the parabola is below the x-axis. Thus $f(0) = c < 0$.

218.

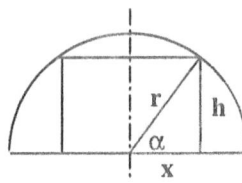

The area of the rectangle $A = 2xh = 2r\cos\alpha \cdot r\sin\alpha = r^2\sin 2\alpha$. $A_{max} = r^2$ at $\sin 2\alpha = 1$, so $a = 45°$. Thus the maximum area of the rectangle can be obtained when $x = h = \dfrac{r\sqrt{2}}{2}$; in other words when the ratio of the sides of a rectangle is 1:2.

219.

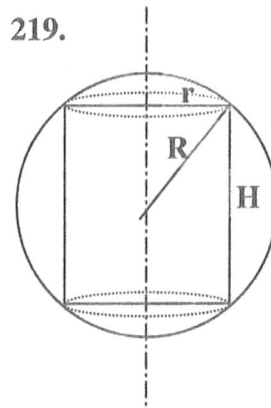

Let R be the radius of the sphere, r and H be the radius and the height of the inscribed cylinder.

The volume of a cylinder $V = \pi r^2 H$. Note that $H = 2\sqrt{R^2 - r^2}$.

Then $V(r) = 2\pi r^2 \cdot \sqrt{R^2 - r^2}$. The derivative of this function is equal to 0 at $r = \dfrac{R\sqrt{6}}{3}$. Then $H = \dfrac{2R\sqrt{3}}{3}$.

220.

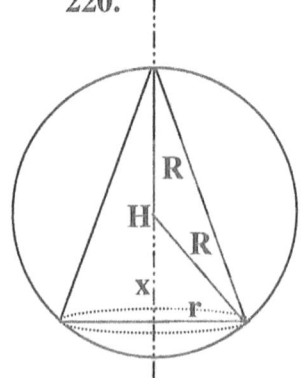

Let R be the radius of the sphere, and let r and b be the radius and the height of an inscribed cone.

The volume of a cone $V = \dfrac{1}{3}\pi r^2 H = \dfrac{1}{3}\pi r^2(R + x) = \dfrac{1}{3}\pi r^2(R + \sqrt{R^2 - r^2})$.

The derivative of this function is $V'(r) = \dfrac{1}{3}\pi \dfrac{2Rr\sqrt{R^2 - r^2} + 2R^2 r - 3r^3}{\sqrt{R^2 - r^2}} = 0$

at $r = \dfrac{2R\sqrt{2}}{3}$. Then $H = \dfrac{4}{3}R$ and $V_{max} = \dfrac{32\pi r^3}{81}$.

221. Let the common zero of both functions be x_0. Then the equation $f(x) - xg(x) = 0$ must have the solution x_0. But $f(x) - xg(x) = 1 - x$, and this equation has the unique solution $x_0 = 1$. $f(1) = a + 2$ and $g(1) = a + 2$. So $f(x)$ and $g(x)$ have a common zero at $a = -2$.

222. Let us find the abscissas of the points of intersection of the two curves.

$y = 2x^2$, $y = a$ $\Rightarrow$ $2x^2 = a$, $x = \pm\sqrt{\dfrac{a}{2}}$. The area between the two curves

$$S = \int_{-\sqrt{\frac{a}{2}}}^{\sqrt{\frac{a}{2}}} (a - 2x^2)dx = \left(ax - \frac{2}{3}x^3\right)\Bigg|_{-\sqrt{\frac{a}{2}}}^{\sqrt{\frac{a}{2}}} = \frac{4}{3}a\sqrt{\frac{a}{2}}; \quad \frac{4}{3}a\sqrt{\frac{a}{2}} = \frac{20\sqrt{30}}{2} \Rightarrow a = 15.$$

223.

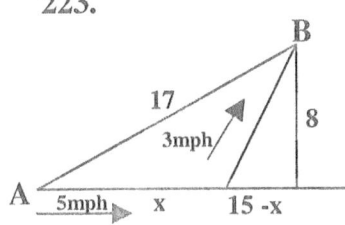

If x is the distance the tourist jogs on the road, then the time it takes him to reach B is $t(x) = \dfrac{x}{5} + \dfrac{\sqrt{8^2 + (15-x)^2}}{3}$. The solution of the equation $t'(x) = 0$ gives the answer $x = 9$ miles.

224. There are 25 pupils in the class. Since 6 pupils in the class have grades of C or D in Math, there are 19 pupils with grades of A, B, or F, and there are at most 19 sportsmen. On the othe hand , if you give each pupil 1 point for practicing a sport, then the number of points is $17 + 13 + 8 = 38$ in the class. Considering that no one practices all three sports, we come to the conclusion that in this class there are precisely 19 sportsmen, and each of them practices two sports. The situation might be shown by the following Venn diagram:

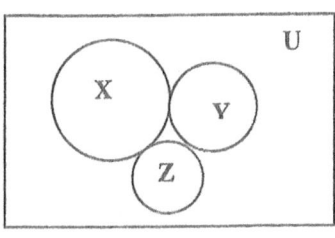

X is the set of sportsmen who are practicing bicycling and swimming, Y is the set of sportsmen who are practicing bicycling and skiing, Z is the set of sportsmen who are practicing swimming and skiing. The set **X** contains 11 sportsmen, the set **Y** contains 6 sportsmen, and the set **Z** contains 2 sportsmen. U is the set of all pupils. Obviously 6 pupils who are not practicing any sport have a C or D, and nobody in this class has an F in Math.

225.

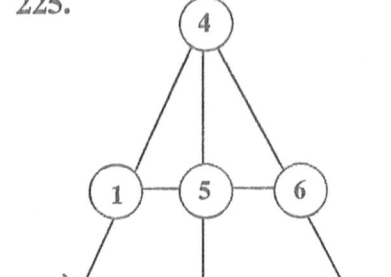

226. The minimum value of the trinomial is equal to 8 at $x = 1$. For the arithmetic sequence $a_1, a_2, a_3, \ldots$ $a_3 + a_9 = 8$ $\Rightarrow$ $a_1 + 2d + a_1 + 8d = 8$ $\Rightarrow$ $2a_1 + 10d = 8$ (d is the difference for the sequence).

The sum of the first 11 terms is $S_{11} = \dfrac{a_1 + a_{11}}{2} \cdot 11 = \dfrac{2a_1 + 10d}{2} \cdot 11 = \dfrac{8}{2} \cdot 11 = 44$.

227. Choosing the balls arbitrarily, in the worst case we will take 9 red, 9 yellow, 9 green and 10 white and black balls. If we now take 1 more ball, then we will have 10 balls of the same color. So we have to take 38 balls.

228. $\log \sin(x + |x|) = 0 \Rightarrow \sin(x + |x|) = 1 \Rightarrow x + |x| = \dfrac{\pi}{2} + 2\pi n$.

If $x > 0$, then $2x = \dfrac{\pi}{2} + 2\pi n$ and $x = \dfrac{\pi}{4} + \pi n$ where $n = 0, 1, 2, 3, \ldots$

If $x \leq 0$, then $x + |x| = 0$, $\sin 0 = 0$ and the equation does not have a solution.

229. 24 miles. (**Hint:** Show that the friends will meet in 4 hours.)

230. Applying the definitions of arithmetic and geometric sequences we have

$$\begin{cases} (x = 2y) - (2x + 3y) = (2x + 3y) - (5x - y) \\ \dfrac{(x-y)^2}{xy+1} = \dfrac{xy+1}{(y+1)^2} \end{cases} \Rightarrow \begin{cases} 2x = 5y \\ (x-1)(y+1) = \pm(xy+1) \end{cases}$$

Now we have two systems to solve :

$$\begin{cases} 2x = 5y \\ (x-1)(y+1) = xy+1 \end{cases} \quad \text{or} \quad \begin{cases} 2x = 5y \\ (x-1)(y+1) = -xy-1 \end{cases}$$

The solution of the first is $x = \dfrac{10}{3}$, $y = \dfrac{4}{3}$. The second equation gives $x = -\dfrac{3}{4}$, $y = -\dfrac{3}{10}$.

231. Let us imagine that each two neighboring dogs are joined by a rope. The approach of each pair of dogs can be modeled as decreasing the length of the rope between them. Obviously the dogs will all meet in the center of the square, when the length of each rope will be zero. Thus the track of each dog is 100 meters. They will meet in 10 seconds. This solution was offered by engineer I.P. Paley.
We reproduce now the alternative explanation by Hugo Steinhaus:

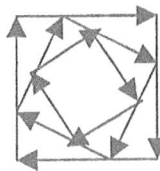

"Since every dog runs at right angles to the direction of the dog which runs after him, and this dog runs directly after him, then the chasing dog approaches his neighbour with the speed of 10 meters per second and will catch him after 10 seconds. Thus the path of each dog is equal to 100 meters. At every moment the four dogs form a square. This square rotates and decreases in area, its sides decrease uniformly at the speed of 10 meters per second. The tracks will intersect at the center S of the original square. They will be curves (logarithmical spirals). They will not intersect earlier, since, if any of the dogs stepped on the track of another dog, it would mean that the other dog had been earlier at the same spot, which is impossible considering that the four distances of the dogs from S are always equal and steadily diminishing."

232. By the definition of an arithmetic sequence ,

$$\log(2^x - 1) = \frac{\log 2 + \log(2^x + 3)}{2} \Rightarrow (2^x - 1)^2 = 2(2^x + 3); \quad 2^{2x} - 4 \cdot 2^x - 5 = 0; \quad 2^x = 5 \text{ or } 2^x = -1.$$

Since the second equation does not have a real solution, the answer is $x = \log_2 5$.

233. By the definition of geometric sequence, $\cos^2\alpha = \dfrac{\sin\alpha \cdot \tan\alpha}{6}$, whence $\cos^2\alpha = \dfrac{\sin^2\alpha}{6\cos\alpha} \Rightarrow$

$6\cos^3\alpha + \cos^2\alpha - 1 = 0$. This equation has only one real solution, $\cos\alpha = \dfrac{1}{2}$.

(Note that $6y^3 + y^2 - 1 = (2y - 1)(3y^2 + 2y + 1)$. You can come to this conclusion after

dividing the trinomial by $(x - \dfrac{1}{2})$.) Thus the three given numbers are the consecutive

terms of a geometric sequence at $\alpha = \pm\dfrac{\pi}{3} + 2\pi n$, where $n = 0, \pm 1, \pm 2, \pm 3, \ldots$.

234. $3^x + 4^x = 5^x \iff \left(\dfrac{3}{5}\right)^x + \left(\dfrac{4}{5}\right)^x = 1$. Applying trigonometry gives us a quick and

interesting result. Let us denote $\sin\alpha = \dfrac{3}{5}$, then $\cos\alpha = \dfrac{4}{5}$ and the simple trigonometric

equation $(\sin\alpha)^x + (\cos\alpha)^x = 1$ makes sense only at $x = 2$.

235. It takes more time to sail by the river than by the lake. Let x be the speed of the
ship in still water and a be the speed of the stream. Note that $x > a > 0$ (in the opposite
case the ship cannot move against the stream). Then to answer the question we have to

prove the inequality $\dfrac{10}{x + a} + \dfrac{10}{x - a} > \dfrac{20}{x}$. The equivalent inequality $\dfrac{x}{x^2 - a^2} - \dfrac{1}{x} > 0$ is

valid for $x > a$. (You can check this by the method of intervals.)

236. Yes. The conclusion is based on solving the equation $130m + 160n = 3000$ for natural
numbers m and n. The workers need to use 12 containers of 130kg and 9 containers
of 160kg.

237. Let us reflect the situation by a Venn Diagram:

X is the set of all blonds, Y is the set of all blue eyed,
Z is the set of the blonds with blue eyes, and
U is the set of all people.

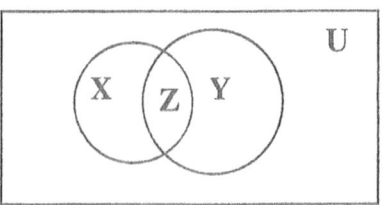

Given $\dfrac{z}{y} > \dfrac{x}{u}$. By the properties of inequalities $\dfrac{z}{x} > \dfrac{y}{u}$.

This means that the part of the blue eyed among the
blonds is more than the part of the blue eyed among
all people.

238. Let the sides of the triangle be a, $a + 1$ and $a + 2$. Obviously side $a + 2$ is the largest
and is opposite the $120°$ angle. By the Cosine Theorem we have

$a + 2 = \sqrt{a^2 + (a + 1)^2 - 2a \cdot (a + 1)^2 \cdot \cos 120°}$.

The solution is $a = 1.5$in. So the triangle has the sides of 1.5in, 2.5in and 3.5in.

239. a) If $n > 0$, then $n = -\log_2\log_2\underbrace{\sqrt{\sqrt{\sqrt{...\sqrt{2}}}}}_{n}$;

b) If $n < 0$, then $n = \log_2\log_2\underbrace{\sqrt{\sqrt{\sqrt{...\sqrt{2}}}}}_{n}$;

c) If $n = 0$, then $n = \log_2\log_2 2$.

240. $n = 6^3 - 6^2 = 216 - 36 = 180$.

241. $n = C_6^3 \cdot C_4^2 \cdot C_2^1 = \dfrac{6 \cdot 5 \cdot 4}{1 \cdot 2 \cdot 3} \cdot \dfrac{4 \cdot 3}{1 \cdot 2} \cdot \dfrac{2}{1} = 240$, where C_m^n is the number of combinations of m things taken n at a time.

242. Let us denote the unknown sides of the triangle by m and n. Then by The Pythagorean Theorem, $m^2 - n^2 = 13^2 = 169$; $(m+n)(m-n) = 169 \cdot 1$, so
$$\begin{cases} m+n = 169 \\ m-n = 1. \end{cases}$$ Answer: $m = 85$, $n = 84$.

243. The members of this club can visit a café either two or three at a time.
If they meet two at a time, we can list all possible combinations :
(1,2), (1,3), (1,4), (2,3), (2,4) ,(3,4).
If 3 people go together, then after this event each of them could meet only with the fourth. So the answer is either 4 or 6 meetings.

244. Answer: 9 masters and 3 grandmasters. (**Hint:** If n is the number of participants, then $\dfrac{n(n-1)}{2}$ is the number of games and the total number of points.)

245. This problem, named "TALKATIVE EVE", is taken from the famous book "Mathematical Circus" by Martin Gardner. We reproduce here his own solution and explanation.
" To solve it, recall that the standard way to obtain the simplest fraction equivalent to a decimal of n repeating digits is to put the repeating period over n 9's and reduce the fraction to its lowest terms." ..." In this instance TALK/9999, reduced to its lowest terms, must equal EVE/DID. DID, consequently, is a factor of 9999. Only three such factors fit DID: 101, 303, 909.
If DID = 101, then EVE /101 = TALK/9999, and EVE = TALK/99. Rearranging terms, TALK = (99) (EVE). EVE cannot be 101 (since we assumed 101 to be DID) and anything larger than 101, when multiplied by 99, has a five-digits product. And so DID = 101 is ruled out.
If DID = 909, then EVE/909 = TALK/9999, and EVE = TALK/11. Rearranging terms, TALK = (11) (EVE). In that case the last digit of TALK would have to be E. Since it is not E, 909 also is ruled out.
Only 303 remains as a possibility for DID. Because EVE must be smaller than 303, E is 1 or 2. Of the 14 possibilities (121, 141, ... , 292) only 242 produces a decimal fitting .TALKTALK..., in which all the digits differ from those in EVE and DID. The unique answer is 242/303 = .798679867986... ."

246. $\dfrac{\sqrt{3}}{1}$

247.
$$\left\{\begin{array}{l} x^2 + y^2 = 2 \\ x^3 + y^3 = 2 \end{array}\right. \Leftrightarrow \left\{\begin{array}{l} (x+y)^2 - 2xy = 2 \\ (x+y)(x^2 - xy + y^2) = 2 \end{array}\right. \Leftrightarrow \left\{\begin{array}{l} (x+y)^2 - 2xy = 2 \\ (x+y)(2-xy) = 2 \end{array}\right.$$

By substituting $u = x + y$ and $v = xy$ we obtain the system
$$\left\{\begin{array}{l} u^2 - 2v = 2 \\ u(2-v) = 2 \,. \end{array}\right.$$

Solving this system by the method of substitution we have
$u_1 = 2,\ v_1 = 1;\ \ u_2 = -1 + \sqrt{3},\ v_2 = 1 - \sqrt{3};\ \ u_3 = -1 - \sqrt{3},\ v_3 = 1 + \sqrt{3}.$
Replacing u by $x + y$ and v by xy and solving the new systems for x and y, we have the following answers:

$$(1,1),\ \left(\frac{\sqrt{3}-1+\sqrt{2\sqrt{3}}}{2},\ \frac{\sqrt{3}-1-\sqrt{2\sqrt{3}}}{2}\right),\ \left(\frac{\sqrt{3}-1-\sqrt{2\sqrt{3}}}{2},\ \frac{\sqrt{3}-1+\sqrt{2\sqrt{3}}}{2}\right).$$

248. Let x be the time in hours for complete burning of the first candle, and t be the time in hours from the lighting of the third candle until the moment when the first and third candles will have the same length. Then 1/x is the part of the first candle that has burned for 1 hour, 1/12 is the part of the second candle that has burned for 1 hour, and 1/8 is the part of the third candle that has burned for 1 hour. Information about the lengths of the candles can be reflected in the following system of equations:
$$\left\{\begin{array}{l} \dfrac{1}{x}(t+1) = \dfrac{1}{8}t \\[2mm] \dfrac{1}{x}(t+3) = \dfrac{1}{12}(t+2). \end{array}\right.$$
Answer: 16 hours.

249.
$$\tan 55° \cdot \tan 65° = \tan(60° - 5°) \cdot \tan(60° + 5°) = \frac{\tan 60° - \tan 5°}{1 - \tan 60° \cdot \tan 5°} \cdot \frac{\tan 60° + \tan 5°}{1 + \tan 60° \cdot \tan 5°} =$$

$$\frac{\tan^2 60° - \tan^2 5°}{1 - \tan^2 60° \cdot \tan^2 5°} = \frac{3 - \tan^2 5°}{1 - 3\tan^2 5°}.$$

$$\tan 5° \cdot \tan 55° \cdot \tan 65° = \tan 5° \cdot \frac{3 - \tan^2 5°}{1 - 3\tan^2 5°} = \tan(3 \cdot 5°) = \tan 15°.$$

$$\tan 5° \cdot \tan 55° \cdot \tan 65° \cdot \tan 75° = \tan 15° \cdot \tan 75° = \tan 15° \cdot \cot 15° = 1.$$

(Note that we used the identity $\tan 3\alpha = \dfrac{3\tan\alpha - \tan^3\alpha}{1 - 3\tan^2\alpha}$ which can easily be proved.)

250. 3, 4, 5.

BIBLIOGRAPHY

1. F. M. Shustef, A. M. Feldman, V. J. Gurevich, "The Set of Olympiad Problems in Mathematics", Minsk, 1962.

2. Hugo Steinhaus with a Foreword by Martin Gardner, "One Hundred Problems in Elementary Mathematics", New York, Basic Books, Inc. 1964.

3. I. H. Sivashinski, "Theorems and Problems in Algebra and Elementary Functions", Moscow, "Science", 1971.

4. M. P. Malaniuk, V.J. Lukavetsky " Olympiads of Young Mathematicians", Kiev, " Radianska Shkola", 1977.

5. Martin Gardner, "Mathematical Circus", New York, Alfred A. Knopf, 1979.

6. O. Ore, " Invitation to the Theory of Numbers", Translation from English, Moscow, "Science", 1980.

7. V. M. Govorov, P. T. Dibov and others, " The Set of Contest Problem in Mathematics", Moscow, " Science", 1983.

8. G. A. Galperin, A. K. Tolpigo, " Moscow Olympiads in Mathematics", Moscow, " Education", 1986.

9. I. S. Petrakov, " Mathematics Clubs for grades 8 - 10", Moscow, " Education", 1987.

10. V. V. Vavilov, I. I. Melnikov and others, "Problems in Mathematics Equations and Inequalities", Moscow, " Science", 1987.

11. Von Lothar Kusch, " Geometry", Cornelsen Druck, Berlin, 1993.

12. M. I. Skanavi, " The Set of Contest Problems in Mathematics", Saint-Petersburg, 1995.

13. L. Altshuler, " Problems to Think", USA, 2012.